EVERYDAY
PROBABILITY
AND STATISTICS

Health, Elections, Gambling and War

2nd Edition

EVERYDAY
PROBABILITY
AND STATISTICS

Health, Elections, Gambling and War

2nd Edition

Michael M. Woolfson

University of York, UK

ICP

Imperial College Press

Published by

Imperial College Press
57 Shelton Street
Covent Garden
London WC2H 9HE

Distributed by

World Scientific Publishing Co. Pte. Ltd.
5 Toh Tuck Link, Singapore 596224
USA office: 27 Warren Street, Suite 401-402, Hackensack, NJ 07601
UK office: 57 Shelton Street, Covent Garden, London WC2H 9HE

British Library Cataloguing-in-Publication Data
A catalogue record for this book is available from the British Library.

EVERYDAY PROBABILITY AND STATISTICS
Health, Elections, Gambling and War
(Second Edition)

ISBN 978-1-84816-761-2
ISBN 978-1-84816-762-9 (pbk)

Typeset by Stallion Press
Email: enquiries@stallionpress.com

Printed by FuIsland Offset Printing (S) Pte Ltd Singapore

Contents

Introduction

There was a time, during the nineteenth century and for much of the twentieth century, when the great and the good in British society would hold that to be considered a "cultured man" one would have to be knowledgeable in Latin, Greek, and English classical literature. Those that held this view, and held themselves to be members of the cultured elite, would often announce, almost with pride, that they had no ability to handle numbers or any mathematical concepts. Indeed, even within living memory, the United Kingdom has had a Chancellor of the Exchequer who claimed that he could only do calculations relating to economics with the aid of matchsticks. Times have changed; society is much more complex than it was a century ago. There is universal suffrage and a greater spread of education. Information is more widely disseminated by newspapers, radio, television and through the internet and there is much greater awareness of the factors that affect our daily lives. The deeds and misdeeds of those that govern, and their strengths and weaknesses, are exposed as never before. It is inevitable that in a multi-party democracy there is a tendency to emphasize, or even overemphasize, the virtues of one's own party and to emphasize, or even overemphasize, the inadequacies of one's opponents.

In this game of claim and counterclaim, affecting education, health, security, social services and all other aspects of national life, statistics plays a dominant role. The judicious use of statistics can be very helpful in making a case and, unfortunately, politicians can rely on the lack of statistical understanding of those they address. An eminent Victorian, and post-Victorian, politician, Leonard Henry Courtney (1832–1913), in a speech about proportional representation, used the phrase *"lies, damned lies and statistics"*[a] and in this he succinctly summarized the role that statistics plays in the hands of those who wish to use it for political advantage. Courtney knew what he was talking about: he was President of the Statistical Society from 1897–1899.

To give a hypothetical example, the party in power wishes to expand a public service. To attract more people in to deliver the service, it has to increase salaries by 20% and in this way it increases the number of personnel by 20%. The party in power say, with pride, that it has invested a great deal of money in the service and that 20% more personnel are delivering it to the nation. The party in opposition say that the nation is getting poor value for money since, for 44% more expenditure, it is getting only 20% more service. Neither party is lying, they are telling neither "lies" nor "damned lies", but they *are* using selective statistics to argue a case.

To make sense of the barrage of numbers with which society is bombarded, one needs to understand what they mean and how they are derived and then one can make informed decisions. Do you want 20% more service for 44% more expenditure, or do you want to retain what service you have with no more expenditure? That is a clear choice and a preference can be made, but the choices cannot be labelled as "bad" or "good" — they are just alternatives.

However, statistics also plays a role in life outside the area of politics. Many medical decisions are made on a statistical basis, since individuals differ in their reactions to medications or surgery in an unpredictable way. In that case, the treatment applied is based on getting the best outcome for as many patients as possible, although some individual patients may not get the best treatment for them.

---

[a] "To My Fellow Disciples at Saratoga Springs", *The National Review* 26 (1895): 21–26.

When resources are limited, then allocating those resources to give the greatest benefit to the greatest number of people may lead to some being denied help that, in principle, it would be possible for them to receive. These are hard decisions that have to be made by politicians and those who run public services. All choices are not in their nature between good and bad — they are often between bad and worse — and a mature society should understand this.

How often have you seen the advertisement that claims that 9 dentists out of 10 recommend toothpaste X? What does it mean? If it means that nine-tenths of all the dentists in the country endorse the product, then that is a formidable claim and one that should give reason to consider changing one's brand of toothpaste. Alternatively, it could mean that, of 10 dentists hand-picked by the company, 9 recommend the toothpaste, and perhaps the dissenting dentist was only chosen to make the claim seem more authentic. Advertisers are adept at making attractive claims for various products, but those claims should be treated with scepticism. Perhaps the antiseptic fluid does kill 99% of all germs but what about the other 1%: are they going to kill you?

Another rich source of manipulated statistics is the press, particularly the so-called tabloid press, the type of newspaper that headlines the antics of an adulterous pop idol, while relegating a major famine in Africa to a small item in an inside page. These newspapers are particularly effective in influencing public opinion and the skilful presentation of selected statistics is often part of this process. In the 1992 UK general election, *The Sun* newspaper, with the largest circulation in the United Kingdom, supported the Conservative Party, and in the few days before the actual vote, presented headlines with no factual or relevant content that were thought to have swayed a significant number of voters. In 1997, *The Sun* switched its allegiance to the Labour Party, which duly won a landslide election victory.

At another level entirely, statistics is the governing factor that controls gambling of any kind: horse racing, card playing, football pools, roulette wheels, dice throwing, Premium Bonds (in the

United Kingdom) and the National Lottery, for example. It is in this area that the public at large seems much more appreciative of the rules of statistics. Many adults who were mediocre, at best, in school mathematics acquire amazing skills, involving an intuitive appreciation of the applications of statistics, when it comes to gambling.

Statistics, as a branch of mathematics, not only has a wide range of applicability, but also has a large number of component topics within it. To master the subject completely requires all the abilities of a professional mathematician, something that is available to comparatively few people. However, to comprehend some of its main ideas and what they mean is, with a little effort, within the capabilities of many people. Here the aim is to explain how statistics impinges on everyday life and to give enough understanding to at least give the reader a fighting chance of detecting when organizations and individuals are trying to pull the wool over the public's eyes. In order to understand statistics one needs also to know something about probability theory, and this too forms a component of this book. In the twenty-first century, a cultured man and woman should understand something of statistics otherwise they will be led by the nose by those that know how to manipulate statistics for their own ends.

More mature readers, who long ago lost contact with formal mathematics, or younger ones who struggle somewhat with the subject, may find it helpful to test their new knowledge by tackling some problems set at the end of each chapter. Worked-out solutions are given so that, even if the reader has not been successful in solving a problem, reading the solution may help to strengthen his or her understanding.

Introduction to Second Edition

Some of the material contained in the first edition of this book, published in 2008, was concerned with decision-making: which horse to back in a race, which medicine to prescribe when the diagnosis is uncertain, or whether or not a country should go to war. In all these decisions some estimates of probabilities are involved, and, in some cases, the estimates are extremely crude and based on intuition and guesswork rather than on hard evidence.

Modern societies worldwide are facing decisions that are of a different order of complexity and importance than have been faced hitherto. These are decisions concerning problems presented to them by the progress of science, some of which may affect the future survival of humankind and even, perhaps, the planet's ability to support any form of life. Scientists, who, in a way, are the initiators of these problems, are also the ones best able to explain the nature of the problems to the lay public, those who, in a democratic society, will, through their politicians, determine the actions to be taken. Many of these problems have become prominent in the recent past. A fierce debate is raging about the question of global warming, with opinions ranging from those who deny that it is a real phenomenon, to those who see a doomsday scenario of the eventual extinction of humankind. Scientists present scientific arguments in support of taking action to reduce the emission of greenhouse gases. Politicians, with their eyes on the next election, are reluctant to take actions that might affect the living standards of their electorate. For the most part, the public at large is confused by the conflicting information they receive and, through natural inertia, are inclined to do nothing.

Another issue of contemporary interest, and also with the potential for catastrophic consequences, is the question of building a new generation of nuclear power stations, suggested as a partial solution to the problem of global warming. This question has been thrown into sharp focus by the disaster at the Japanese nuclear power station at Fukushima in March 2011, second in scale only to the Chernobyl disaster of April 1986.

What can scientists and engineers tell us about the safety of nuclear power generation and the probabilities of various scenarios for global warming, and what confidence can we have in what they say? It is all a matter of probabilities and confidence: a large part of what this book is about. To illustrate the problem of bridging the gap between scientists and the general public, the topic of modern meteorology is described, where it is pointed out that, despite the public's desire for certain and unequivocal forecasts, this is something that science cannot deliver.

A final topic introduced in the second edition, which is causing a considerable stir at the beginning of the twenty-first century, is that of providing pensions for a population of ever increasing life expectancy. With the change of age distribution, with a higher proportion of older people in society, the scale of provision made by pension funds in the past can no longer be financially sustained, and an increasing number of funds are running into deficit. The nature of the problem, and what is required to solve it, is explained — but there is no soft option.

24th May 2011

Chapter 1

The Nature of Probability

Probable impossibilities are to be preferred to improbable possibilities
(Aristotle, 350 BCE, *Poetics*)

1.1. Probability and Everyday Speech

The life experienced by any individual consists of a series of events within which he or she plays a central role. Some of these events, like the rising and setting of the sun, occur without fail each day. Others occur often, sometimes on a regular, if not daily, basis and might, or might not, be predictable. For example, going to work is normally a predictable and frequent event but the mishaps, such as illnesses, that occasionally prevent someone from going to work are events that are to be expected from time to time but can be predicted neither in frequency nor timing. To the extent that we can, we try to compensate for the undesirable uncertainties of life — by making sure that our homes are reasonably secure against burglary, a comparatively rare event despite public perception — or by taking out insurance against contingencies, such as loss of income due to ill health or car accidents.

To express the likelihood of the various events that define and govern our lives we have available a battery of words with different shades of meaning, some of which are virtually synonymous. Most of these words are so basic that they can best be defined in terms of each other. If we say that something is *certain* then we mean that the event will happen without a shadow of doubt: on any day, outside the polar regions, we are certain that the sun will set. We can qualify *certain* with an adverb by saying that something is *almost certain* meaning that there is only a very small likelihood that it would not happen. It is almost certain that rain will fall sometime during next January because that month and February are the wettest months of the year in the United Kingdom. There are rare years when it does not rain in January but these represent freak conditions. However, when we say that an event is *likely*, or *probable*, we imply that the chance of it happening is greater than it not happening. August is usually sunny and warm and it is not unusual for there to be no rain in that month. Nevertheless, it is probable that there will be some rain in August because that happens in most years.

The word *possible* or *feasible* could just mean that an event is capable of happening without any connotation of likelihood, but in some contexts it could be taken to mean that the likelihood is not very great, or that the event is *unlikely*. Finally, *impossible* is a word without any ambiguity of meaning; the event is incapable of happening under any circumstances. By attaching various qualifiers to these words — *almost impossible* as an example — we can obtain a panoply of overlapping meanings, but at the end of the day, with the exception of the extreme words, certain and impossible, there is a subjective element in both their usage and interpretation.

While these fuzzy descriptions of the likelihood that events might occur may serve in everyday life, they are clearly unsuitable for scientific use. Something much more objective, and numerically defined, is needed.

1.2. Spinning a Coin

We are all familiar with the action of spinning a coin: it happens at cricket matches to decide which team chooses who will bat first and at football matches to decide which team can choose the end of the pitch to play in the first half. There are three possible outcomes to the event of spinning a coin: head, tail or standing on an edge. That comes from the shape of a coin, which is a thin disk (Fig. 1.1).

Fig. 1.1.   The three possible outcomes for spinning a coin.

However, the shape of the coin contains another element, that of symmetry. Discounting the possibility that the coin will end up standing on one edge (unlikely but feasible in the general language of probabilities), we deduce from symmetry that the probability of a head facing upwards is the same as that for a tail facing up. If we were to spin a coin 100 times and we obtained a head each time, we would suspect that something was wrong — either that it was a trick coin with a head on each side, or one that was so heavily biased it could only come down one way. From an instinctive feeling of the symmetry of the event, we would expect that the two outcomes had equal probability, so that the most likely result of spinning the coin 100 times would be 50 heads and 50 tails, or something fairly close to that result. Since we expect a tail 50% of the times we spin the coin, we say that the probability of getting a tail is $\frac{1}{2}$, because that is the fraction of the occasions that we expect that outcome. Similarly, the probability of getting a head is $\frac{1}{2}$. We have taken the first step in assigning a numerical value to the likelihood, or probability, of the occurrence of particular outcomes.

Supposing that we repeated the above experiment of spinning a coin but this time it was with the trick coin (the one with a head on both sides). Every time we spin the coin we get a head; it happens 100% of the time. We now say that the probability of getting a head is 1 because that is the fraction of the occasions we expect that outcome. Getting a head is certain and that is what is meant by a probability of 1. Conversely, we get a tail on 0% of the times we flip the coin; the probability of getting a tail is 0. Getting a tail is impossible and

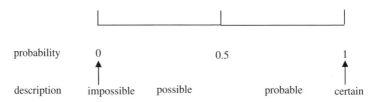

Fig. 1.2. The numerical probability range with some notional verbal descriptions of regions.

that is what is meant by a probability of 0. Figure 1.2 shows this assignment of probabilities in a graphical way.

The range shown for probability in Fig. 1.2 is complete. A probability cannot be greater than 1 because no event can be more certain than certain. Similarly, no probability can be less than 0, i.e., negative, since no event can be less possible than impossible.

We are now in a position to express the probabilities for spinning an unbiased coin in a mathematical form. If the probabilities of getting a head or a tail are p_h and p_t respectively, then we can write:

$$p_h = p_t = \frac{1}{2}. \tag{1.1}$$

1.3.   Throwing or Spinning Other Objects

Discounting the slight possibility of it standing on edge, there are just two possible outcomes of spinning a coin, head or tail. This is something that comes from the symmetry of a disk. However, if we throw a die, then there are six possible outcomes: 1, 2, 3, 4, 5 or 6. A die is a cubic object with six faces and, without numbers marked on them, all the faces are similar and similarly disposed with respect to other faces (Fig. 1.3).

From the symmetry of the die, it would be expected that the fraction of throws yielding a particular number, say a 4, would be $\frac{1}{6}$, so that the probability of getting a 4 is $p_4 = \frac{1}{6}$ and that would be the same probability of getting any other specified number. Analogous

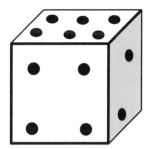

Fig. 1.3. A die showing three of the six faces.

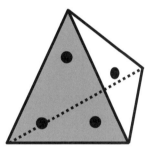

Fig. 1.4. A regular tetrahedron — a "die" with four equal-probability outcomes.

to the coin equation (1.1), we have for the probability of each of the six possible outcomes:

$$p_1 = p_2 = p_3 = p_4 = p_5 = p_6 = \frac{1}{6}. \tag{1.2}$$

It is possible to produce other symmetrical objects that would give other numbers of possible outcomes, each with the same probability. In Fig. 1.4 we see a regular tetrahedron, a solid object with four faces each of which is an equilateral triangle. The two faces we cannot see have two and four spots on them, respectively. This object would not tumble very well if thrown onto a flat surface unless thrown quite violently but, in principle, it would give, with equal probability, the numbers 1–4 so that:

$$p_1 = p_2 = p_3 = p_4 = \frac{1}{4}. \tag{1.3}$$

Fig. 1.5. A device for giving $p_1 = p_2 = p_3 = p_4 = p_5 = \frac{1}{5}$.

A better device in terms of its ease of use is a regular-shaped polygon mounted on a spindle about which it can be spun. This is shown in Fig. 1.5 for a device giving numbers 1–5 with equal probability. The spindle is through the centre of the pentagon and perpendicular to it. The pentagon is spun about the spindle axis like a top and eventually comes to rest with one of the straight boundary edges resting on the supporting surface, which indicated the number for that spin.

We have now been introduced to the idea of probability expressed as a fractional number between 0 and 1: the only useful way for a scientist or mathematician. Next we will consider slightly more complicated aspects of probability when combinations of different outcomes can occur.

Problems 1

1.1 Meteorology is not an exact science and hence weather forecasts have to be couched in terms that express that lack of precision. The following is a Meteorological Office forecast for the United Kingdom covering the period 23rd September to 2nd October 2006:

> *Low pressure is expected to affect northern and western parts of the UK throughout the period. There is a risk of some showery rain over south-eastern parts over the first weekend but otherwise much of eastern England and possibly eastern Scotland*

should be fine. More central and western parts of the UK are likely to be rather unsettled with showers and some spells of rain at times, along with some periods of strong winds too. However, with a southerly airflow dominating, rather warm conditions are expected with warm weather in any sunshine in the east.

Identify all those parts of this report that indicate lack of certainty.

1.2 The figure below is a dodecahedron. It has twelve faces, each a regular pentagon, and each face is similarly disposed with respect to the other 11 faces. If the faces are marked with the numbers 1 to 12, then what is the probability of getting a 6 if the dodecahedron is thrown?

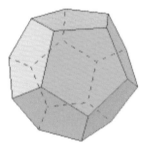

1.3 The object shown below is a truncated cone, i.e., a cone with the top sliced off with a cut parallel to the base.

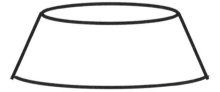

Make drawings showing the possible ways that the object can come to rest if it is thrown onto the floor. Based on your intuition, which will be the most and least probable ways for the object to come to rest?

1.4 A certain disease can be fatal and it is known that 123 out of 4,205 patients in a recent epidemic died. As deduced from this information, what is the probability, expressed to three decimal places, that a given patient contracting the disease will die?

Chapter 2

Combining Probabilities

But to us, probability is the very guide of life (Bishop Joseph Butler, 1726, *Fifteen Sermons Preached at the Rolls Chapel, Botham*)

2.1. Either–or Probability

Let us consider the situation where a die is thrown, and we wish to know the probability that the outcome will be *either* a 1 *or* a 6. How do we find this? First, we consider the six possible outcomes, all of equal probability. Two of these, a 1 and a 6 — one third of the possible outcomes — satisfy our requirement, so the probability of obtaining a 1 or a 6 is:

$$p_{1\,or\,6} = \frac{2}{6} = \frac{1}{3}. \tag{2.1}$$

Another way of expressing this result is to say that:

$$p_{1\,or\,6} = p_1 + p_6 = \frac{1}{6} + \frac{1}{6} = \frac{1}{3}. \tag{2.2}$$

In words, Equation (2.2) says that "the probability of getting either a 1 or a 6 is the sum of the probabilities of getting a 1 and getting a 6".

This idea can be extended so that the probability of getting one of 1, 2 or 3 when throwing the die is:

$$p_{1\,or\,2\,or\,3} = p_1 + p_2 + p_3 = \frac{1}{6} + \frac{1}{6} + \frac{1}{6} = \frac{1}{2}. \tag{2.3}$$

Similarly, the probability of getting either a head or a tail when flipping a coin is:

$$p_{h\,or\,t} = p_h + p_t = \frac{1}{2} + \frac{1}{2} = 1, \qquad (2.4)$$

which corresponds to certainty, since the only possible outcomes are either a head or a tail.

In considering these combinations of probability we are taking alternative outcomes of a single event, for example, throwing a die. If we are interested in the outcome being a 1, 2 or 3 then, if we obtain a 1, we exclude the possibility of having obtained either of the other outcomes of interest: a 2 or a 3 (Fig. 2.1). Similarly, if we obtained a 2, the outcomes 1 and 3 would have been excluded. The outcomes for which the probabilities are being combined are *mutually exclusive*. It is a general rule that *the probability of having an outcome that is one or other of a set of mutually exclusive outcomes is the sum of the probabilities for each of them taken separately.*

To explore this idea further consider a standard pack of 52 cards. The probability of choosing a particular card by a random selection is $\frac{1}{52}$. Four of the cards are jacks so the probability of picking a jack is:

$$p_J = \frac{1}{52} + \frac{1}{52} + \frac{1}{52} + \frac{1}{52} = \frac{4}{52} = \frac{1}{13}. \qquad (2.5)$$

Now we want to know the probability of picking a court card (i.e., jack, queen or king) from the pack. The separate probabilities

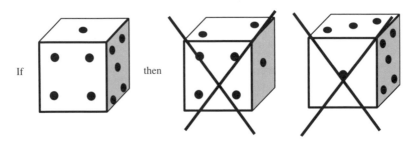

Fig. 2.1.   In either–or probability if a 1 is obtained then a 2 or a 3 is excluded.

of outcomes for jacks, queens and kings are all $\frac{1}{13}$ but the outcomes of obtaining a jack, a queen or a king are mutually exclusive. Hence the probability of picking a court card is:

$$p_{\text{J or Q or K}} = p_\text{J} + p_\text{Q} + p_\text{K} = \frac{1}{13} + \frac{1}{13} + \frac{1}{13} = \frac{3}{13}. \qquad (2.6)$$

Of course, one could have lumped court cards together as a single category, and since there are 12 of them in a pack of 52, the probability of selecting one of them could have been found directly as: $\frac{12}{52} = \frac{3}{13}$. However, having a mathematically formal way of considering problems is sometimes helpful in less obvious cases.

This kind of combination of probabilities has been called *either–or* and a pedantic interpretation of the English language would give the inference that only two possible outcomes could be involved. However, that is not a mathematical restriction and this type of probability combination can be applied to any number of mutually exclusive outcomes.

2.2.   Both–and Probability

Now we imagine that two events occur: a coin is spun and a die is thrown. We now ask the question "What is the probability that we get *both* a head *and* a 6?" The two outcomes are certainly not mutually exclusive — indeed they are *independent* outcomes. The result of spinning the coin can have no conceivable influence on the result of throwing the die. First we list all the outcomes that are possible:

$$\begin{array}{cccccc} h+1 & h+2 & h+3 & h+4 & h+5 & h+6 \\ t+1 & t+2 & t+3 & t+4 & t+5 & t+6 \end{array}$$

There are 12 possible outcomes, each of equal probability, and we are concerned with the one marked with an arrow. Clearly, the probability of having *both* a head *and* a 6 is $\frac{1}{12}$. This probability can be considered in two stages. First we consider the probability of obtaining a head, which is $\frac{1}{2}$. This corresponds to the outcomes on the top row of our list. Now we consider the probability that we also have a 6 that restricts us to 1 in 6 of the combinations in the top row, since the probability of getting a 6 is $\frac{1}{6}$. Looked at in two stages we see that:

$$p_{\text{both h and 6}} = p_{\text{h}} \times p_6 = \frac{1}{2} \times \frac{1}{6} = \frac{1}{12}. \qquad (2.7)$$

This rule can be extended to find the combined probability of any number of *independent* events. Thus, the combined probability that spinning a coin gives a head, throwing a die gives a 6 and picking a card from a pack gives a jack is:

$$p_{\text{h and 6 and J}} = p_{\text{h}} \times p_6 \times p_J = \frac{1}{2} \times \frac{1}{6} \times \frac{1}{13} = \frac{1}{156}. \qquad (2.8)$$

Once again, our description of this probability combination has done violence to the English language. The combination *both–and* should formally be applied only to two outcomes but we stretch it to describe the combination of any number of independent outcomes.

These rules of "either–or" and "both–and" combinations of probability can themselves be combined to solve quite complicated probability problems.

2.3.   Genetically Inherited Disease — Just Gene Dependent

Within any population there will exist a number of genetically inherited diseases. There are about 4,000 such diseases known and particular diseases tend to be prevalent in particular ethnic groups. For example, sickle cell anaemia is mainly present in people of West African origin, which will include many of the black populations of the Caribbean and North America and also of the United Kingdom. This disease affects the haemoglobin molecules contained within red blood cells, which are responsible for carrying oxygen from the lungs to muscles in the body and carbon dioxide back from the muscles to the lungs. The haemoglobin forms long rods within the cells, distorting them into a sickle shape and making them less flexible so that they flow less easily. In addition, the cells live less time than the normal 120 days for a healthy red cell and so the patient suffers from a constant state of anaemia. Another genetic disease is Tay–Sachs disease that affects people of Jewish origin. This attacks the nervous system, destroying brain and nerve cells, and is always fatal, usually at the infant stage.

To understand how genetically-transmitted diseases are transmitted we need to know something about the gene structure of living matter, including humans. Contained within each cell of a human being there are a large number of chromosomes, thread-like bodies which contain, strung out along them, large numbers of genes. The number of genes controlling human characteristics is somewhere in the range of 30,000 to 40,000. Each gene is a chain of DNA of length anywhere from 1,000 to hundreds of thousands of the base units that make up DNA. Genes occur in pairs which usually correspond to contrasting hereditary characteristics. For example, one gene pair might control stature so that gene A predisposes towards height, while the other member of the pair, gene a, gives a tendency to produce shorter individuals. Each person has two of these stature genes in their cells. If they are both A then the person will have a tendency to be tall and if they are both a then there will be a tendency to be short. One can only talk about tendency in this instance since other factors influence stature, in particular diet. A child inherits one "stature" gene from each parent and which of the two genes they get from each parent is purely random. Here we show some of the possibilities for various parental contributions:

Father	Mother	Child (all pairs of equal probability)			
Aa	*Aa*	*AA*	*Aa*	*aA*	*aa*
AA	*aa*	*Aa*	*Aa*	*Aa*	*Aa*
AA	*AA*	*AA*	*AA*	*AA*	*AA*

When an individual has a contrasting gene pair then sometimes the characteristics will combine, so that, for example, *Aa* will tend to give a medium-height individual, but in other cases one of the genes may be dominant. Thus if *B* is a "brown-eye" gene and *b* is a "blue-eye" gene, then *BB* gives an individual with brown eyes, *bb* gives an individual with blue eyes, and *Bb* (equivalent to *bB*) will give brown eyes because *B* is the dominant gene. Sometimes genes become "fixed" in a population. All Chinese are *BB*, so that all Chinese children must inherit the gene pair *BB* from their parents. Thus all Chinese have brown eyes.

Now we consider a genetically-related disease linked to the gene pair *Dd*. The gene *d* predisposes towards the disease and someone who inherits a pair *dd* will certainly get the disease and die before maturity. However, we take it that *d* is a very rare gene in the community and *D* is dominant. Anyone who happens to be *Dd* will be free of the disease but may pass on the harmful gene *d* to his or her children; such a person is known as a *carrier*. Let us suppose that in this particular population the ratio of *d:D* is 1:100. What is the probability that, with random mating, i.e., no monitoring of parents, a baby born in that population will have the disease?

We can consider this problem by considering the allocation of the gene pair to the baby one at a time. The probability that the first gene will be *d* is 0.01 because that is the proportion of the *d* gene. The allocation of the second gene of the pair is independent of what the first one happens to be so, again, the probability that this one is *d* is 0.01. Hence the probability that *both* the first gene is *d* *and* the second gene is *d* is $0.01 \times 0.01 = 0.0001$, or 1 chance in 10,000. If we were interested in how many babies would be *carriers* of the faulty gene, i.e., possessing the gene pair *Dd*, then, using both–and probability, we note that:

the probability that *both* gene 1 is *D* *and* gene 2 is *d* is
$0.99 \times 0.01 = 0.0099$,
the probability that *both* gene 1 is *d* *and* gene 2 is *D* is
$0.01 \times 0.99 = 0.0099$.

Since Dd and dD are mutually exclusive the probability of the baby carrying the gene pairs *either Dd or dD* is:

$$0.0099 + 0.0099 = 0.0198,$$

so that about 1 in 50 babies born is a carrier.

Some genetically-related diseases are very rare indeed because the incidence of the flawed gene is low; for $d:D = 1{:}1{,}000$ only 1 child in 1,000,000 would contract the disease although about 1 in 500 of them would be carriers of the disease. On the other hand, diabetes, which is thought to have a genetically-transmitted element, is much more common and the number of carriers is probably quite high.

2.4.   Genetically Inherited Disease — Gender Dependent

There are genetically-transmitted diseases where the gender of the individual is an important factor. The sex of an individual is determined by two chromosomes, X and Y, a female having the chromosome pair XX and a male XY. The female always contributes the same chromosome to her offspring, X, but the male contributes X or Y with equal probability, thus giving a balance between the numbers of males and females in the population. The English king Henry VIII divorced two wives because they did not give him a son but now we know that he was to blame for this! There are some defective genes, for example, that which leads to haemophilia, which only occurs in the X chromosome. If we call a chromosome with the defective X gene X' then the following situations can occur:

Daughter with the chromosome combination XX' will be a carrier of haemophilia but will not have the disease because the presence of X compensates for the X'.

Son with combination $X'Y$ will have the disease because there is no accompanying X to compensate for the X'.

Now we can see various outcomes from different parental chromosome compositions:

(i)

All children free of haemophilia and none are carriers.

(ii)

The father suffers from the disease but all children are free
of haemophilia. Sons are completely unaffected because they just
receive a Y chromosome from their fathers. However, all daughters
are XX' and so all are carriers.

(iii)

Here the father is free of the disease but we have a carrier
mother. Half the sons are $X'Y$ and so are haemophiliacs. The other
half of the sons are XY and so are free of the disease. Half of the
daughters are XX' and so are carriers but the other half are XX and
are not carriers.

(iv)

Here we have a haemophiliac father with a carrier mother. Half
the sons are $X'Y$ and so are haemophiliacs. The other half of the sons
are XY and so are free of the disease. Half the daughters are $X'X'$
and so will have the disease while the other half are $X'X$ and so are
carriers.

The incidence of haemophilia does not seem to have any strong correlation with ethnicity and about 1 in 5,000 males are born with the disease. The most famous case of a family history of haemophilia is concerned with Queen Victoria, who is believed to have been a carrier. She had 9 children and 40 grandchildren and several male descendants in royal houses throughout Europe suffered from the disease. The most notable example was that of Tsarevich Alexis, the long-awaited male heir to the throne of Russia, born in 1904. A coarse and rather dissolute priest, Rasputin, became influential in the Russian court because of his apparent ability to ameliorate the symptoms of the disease in the young Alexis. There are many who believe that his malign influence on the court was an important contributory factor that led to the Russian revolution of 1917.

2.5.   A Dice Game — American Craps

American craps is a dice game that is said to originate from Roman times and was probably introduced to America by French colonists. It involves throwing two dice and the progress of the game depends on the sum of the numbers on the two faces. If the thrower gets a sum of either 7 or 11 in their first throw that is called "a natural" and they immediately win the game. However, if they get 2 (known as "snake-eyes"), 3 or 12 they immediately lose (Fig. 2.2). Any other sum is the players "point". The player then continues throwing until either they get their "point" again, in which case they win, or until they get a 7, in which case they lose. The probabilities of obtaining various sums are clearly the essence of this game.

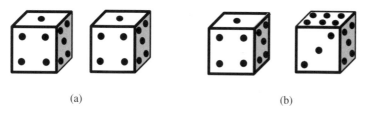

(a)   (b)

Fig. 2.2.   (a) Snake-eyes; (b) A *natural* (first throw) but a losing throw later.

We first consider the probability of obtaining 2, 3 or 12. The two dice must show one of:

$$1+1 \quad 1+2 \quad 2+1 \quad 6+6$$

and these are mutually exclusive combinations. Since the numbers on each of the dice are independent of each other, each combination has a probability of $\frac{1}{36}$, for example:

$$p_{\text{snake-eyes}} = p_1 \times p_1 = \frac{1}{6} \times \frac{1}{6} = \frac{1}{36}. \tag{2.9}$$

Hence the probability of getting 2, 3 or 12 is:

$$p_{2,3\,\text{or}\,12} = p_{1+1} + p_{1+2} + p_{2+1} + p_{6+6} = \frac{4}{36} = \frac{1}{9}. \tag{2.10}$$

Getting 7 or 11 as a sum requires one or other of the combinations:

$$6+1 \quad 5+2 \quad 4+3 \quad 3+4 \quad 2+5 \quad 1+6 \quad 6+5 \quad 5+6$$

These combinations are mutually exclusive, each with a probability of $\frac{1}{36}$ so that the probability of a "natural" is $\frac{8}{36} = \frac{2}{9}$.

If the player has to play to their "point" then their chances of winning depends on the value of that "point" and how the probability of obtaining it compares with the probability of obtaining 7. Below, we show for each possible "point" the combinations that can give it and the probability of achieving it per throw:

Point	Combinations					Probability
4	$3+1$	$2+2$	$1+3$			$\frac{3}{36} = \frac{1}{12}$
5	$4+1$	$3+2$	$2+3$	$1+4$		$\frac{4}{36} = \frac{1}{9}$
6	$5+1$	$4+2$	$3+3$	$2+4$	$1+5$	$\frac{5}{36}$
8	$6+2$	$5+3$	$4+4$	$3+5$	$2+5$	$\frac{5}{36}$
9	$6+3$	$5+4$	$4+5$	$3+6$		$\frac{4}{36} = \frac{1}{9}$
10	$6+4$	$5+5$	$4+6$			$\frac{3}{36} = \frac{1}{12}$

The total of 7 can be achieved in 6 possible ways (see above ways of obtaining a "natural") and its probability of attainment, $\frac{1}{6}$, is higher than that of any of the points.

This is an extremely popular gambling game in the United States and those that play it almost certainly acquire a good instinctive feel for probability theory as it affects this game, although few would be able to express their instinctive knowledge in a formal mathematical way. Later, when we know a little more about probabilities, we shall calculate the overall chance of the thrower winning — it turns out that the odds are very slightly against him.

Problems 2

2.1 For the dodecahedron described in problem 1.2, what is the probability that a throw will give a number less than 6?

2.2 If a card is drawn from a pack of cards, what is the probability that it is an ace?

2.3 The dodecahedron from problem 1.2 and a normal six-sided die are thrown together. What is the probability that the both the dodecahedron and the die will give a 5?

2.4 Find the combined probability of drawing a jack from a pack of cards and getting four heads from spinning four coins. Is this probability greater than or less than that found in problem 2.3?

2.5 The dodecahedron from problem 1.2 and a normal six-sided die are thrown together. What is the probability that the sum of the numbers on the sides is 6?

(*Hint*: Consider the number of ways that a total of six can be obtained and the probability of each of those ways.)

2.6 For a particular gene pair F and f, the latter gene leads to a particular disease, although F is the dominant gene. If the ratio of the incidence of the genes is $f: F = 1 : 40$, then what proportion of the population is expected to be:

(i) affected by the disease?

(ii) carriers of the disease?

Chapter 3

A Day at the Races

Do not trust the horse (Virgil, 19 BCE, *The Aeneid*)

3.1. Kinds of Probability

The kind of probability we have mostly considered numerically thus far can be regarded as *logical probability*, which is to say that we can estimate probability by an exercise of logic, or reason. Usually, this is based on the concept of symmetry; a coin has two sides, neither of which is special in a probabilistic sense, so each has an associated probability of $\frac{1}{2}$. A die is a symmetrical object, each face of which is related geometrically in the same way to the other five faces. Here, again, we divide the probability 1, corresponding to certainty, into six equal parts and associate a probability of $\frac{1}{6}$ with each of the faces. Such assessments of probability are both logical and intuitive.

If we wish to know the likelihood that an infected person will die from a potentially fatal disease, we cannot use logical probability. Here the only guide is previous experience, i.e., knowledge of the death rate from the disease, assessed from those that have contracted it previously in similar circumstances. That was the basis of Problem 1.4. This kind of probability, gained from observation or, where appropriate, from experiment, is known as *empirical probability* and it is the more important kind of probability from the point of view of everyday life — unless one is an inveterate gambler in casinos, where many games are based on logical probabilities. Thus, if one is planning a two-week holiday in Florida, one could predict the probability of being within 10 miles of a tornado during the holiday by consulting the weather statistics for that state. The total area within 10 miles of any spot is the area of a circle of that radius and is:

$$A = \pi \times (10)^2 \text{ square miles} = 314 \text{ square miles}.$$

The average frequency of tornadoes in Florida is 9.09 per year, per 10,000 square miles. So, assuming that all times of the year have the same probability, the number expected within an area of 314 square miles in a particular two-week period is:

$$N = 9.09 \times \frac{314}{10,000} \times \frac{2}{52} = 0.011.$$

Of course, fractions of a tornado do not exist, but what the answer means is that the probability of having a tornado within a distance of 10 miles is 0.011, or about 1 in 90. For a tornado to be a risk to an individual they would have to be in its direct path and the risk of that is much smaller than 1 in 90: probably more like 1 in 30,000, so it is very small.

A kind of probability, which is empirical in essence but has an extra dimension, is in assessing the likelihood that a horse, or perhaps a greyhound, will win a particular race. There is, of course, the previous record of the contesting animals but there are many other factors to take into account. Each of the previous races, on which the form would be assessed, would have taken place under different conditions, both in terms of the state of the track and the characteristics of the opposition. In horse races like the Grand National (held at Aintree, Liverpool, UK), which involves horses jumping over high fences, the unpredictable interaction of horses as they jump together over a fence can cause upsets in the race result. Again, like humans, horses have their good days and bad days and these cannot be predicted. The kind of probability involved here is certainly neither logical nor even empirical, since each case being considered has its own individual features and cannot be accurately assessed from past performance. The extra dimension that must be involved here in assessing probabilities is judgement, and we can define this kind of probability assessment as *judgemental probability*. Getting this right is of critical importance to bookmakers, who take bets on the outcome of races and who could be ruined by consistently making faulty judgements.

3.2. Betting on a Horse

There are three main ways that a punter can bet on the results of a race: in a betting shop, usually locally, or at the racetrack, through a bookmaker or the Tote. The UK Tote is a system that, by its very nature, cannot lose money. At the time a punter places a £10 bet on a horse, called, say, My Fair Lady, with the Tote, he has only a rough idea of what he would win if that horse won its race. The Tote organizers will take the total money staked on the race, say, £500,000, take a proportion of this, say, £25,000, to cover expenses and create a profit, and then divide the remaining money (£475,000), between those that backed My Fair Lady to win in proportion to how much they staked. Thus, if the people backing My Fair Lady staked a total of £20,000 on the horse, then our punter would receive a fraction (10/20,000) of the returned stake money. Thus, his return will be:

$$£475,000 \times \frac{10}{20,000} = £237.50.$$

The Tote is actually a little more complicated than we have described because "place bets" can be made where the punter will get a reduced return (a fraction of the odds for winning), if the horse finishes in one of the first two, three or four places, depending on the number of horses in the race. However, the essentials of the Tote system are described by just considering the bets to win.

A betting shop or a bookmaker, by contrast, enters into a contract with the punter to pay particular odds at the time the bet is placed. If the horse backed has odds of 3/1 (three to one), then a successful £10 bet will bring a return of the original £10, plus 3 times £10, i.e., £40 in all. A bookmaker can lose, and lose heavily, on a particular race but if he knows his craft and assesses the odds skilfully, he will make a profit over the long term.

Let us take an example of odds set by a bookmaker that would obviously be foolish. Consider a two-horse race, where the bookmaker sets odds of 2/1 on each horse. A punter would not take long to work out that if he staked £1,000 on each horse his total stake on the race would be £2,000 but, whichever horse won, his return from the race would be £3,000. This case is easy to see without analysing it in detail but more subtle examples of the bad setting of odds can occur. Let us take a hypothetical race with 10 runners, where the bookmaker, assessing the relative merits of the various runners, offers the following odds on a race for fillies:

Diana	3/1
Dawn Lady	6/1
Fairy Princess	6/1
Olive Green	10/1
Mayfly	10/1
Dawn Chorus	15/1
Missy	15/1
Lovelorn	20/1
Helen of Troy	25/1
Piece of Cake	30/1

Although it is not obvious from a casual inspection of the list of odds, this bookmaker is heading for certain ruin and any punter worth his salt could make a profit on this race. Let us see how he does this. He bets on every horse and places bets as follows:

Diana	£250	(£1,000)
Dawn Lady	£143	(£1,001)
Fairy Princess	£143	(£1,001)
Olive Green	£91	(£1,001)

Mayfly	£91	(£1,001)
Dawn Chorus	£63	(£1,008)
Missy	£63	(£1,008)
Lovelorn	£48	(£1,008)
Helen of Troy	£39	(£1,014)
Piece of Cake	£33	(£1,023)

In parentheses, after the bets, we see what the punter will receive, if that particular horse wins; this is anything between £1,000 and £1,023. However, if the bets are added they come to £964; the punter is a certain winner and he will win between £36 and £59, depending on which horse wins the race. Clearly, no bookmaker would offer such odds on a race and we shall now see what the principles are for setting the odds that ensure a high probability of profit for the bookmaker in the longer-term.

We notice that what the punter has done is to set his stake on each horse at a whole number of pounds that will give a return (stake money plus winnings) of £1,000, or a small amount more. For simplicity in analysing the situation, in what follows we shall assume that he adjusts his stake to receive exactly £1,000, even though bookmakers do not accept stakes involving pennies and fractions of pennies. If the horse has odds of $n/1$, then the punter will receive $n + 1$ times his stake money, so, to get a return of £1,000, the amount staked is: £1,000/$(n + 1)$. To check this we see that for Diana, with $n = 3$, the stake is £1,000/$(3 + 1)$ = £250. Now let us take a general case, where the odds for a ten-horse race are indicated as: $n_1/1, n_2/1 \ldots n_{10}/1$. If the punter backs every horse in the race, planning to get £1,000 returned no matter which horse wins, then his total stake, in pounds, is:

$$S = \frac{1,000}{n_1 + 1} + \frac{1,000}{n_2 + 1} + \cdots + \frac{1,000}{n_{10} + 1}. \tag{3.1}$$

If S is less than £1,000, then the punter is bound to win. Thus, for the bookmaker not to be certain to lose money to the knowledgeable punter, S must be greater than 1,000. This condition gives the book-maker's golden rule, which we will now express in a mathematical form. If we take Equation (3.1) and divide both sides by 1,000, then we have:

$$\frac{S}{1,000} = \frac{1}{n_1 + 1} + \frac{1}{n_2 + 1} + \cdots + \frac{1}{n_{10} + 1} \tag{3.2}$$

and for the bookmaker not to lose to the clever punter, the left-hand side must be greater than 1. Put into a mathematical notation, which we first give and then explain, the golden rule condition is:

$$\sum_{i=1}^{m} \frac{1}{n_i + 1} > 1. \tag{3.3}$$

The symbol ">" means "greater than" and Equation (3.3) is the condition that a punter cannot *guarantee* to win, although, of course, he may win just by putting a single bet on the winning horse. The summation symbol $\sum_{i=1}^{m}$ means that you are going to add together m quantities with the value of i going from 1 to m. In our specific case $m = 10$, the number of horses, and i runs from 1 to 10 so that:

$$\sum_{i=1}^{10} \frac{1}{n_i + 1} = \frac{1}{n_1 + 1} + \frac{1}{n_2 + 1} + \frac{1}{n_3 + 1} + \frac{1}{n_4 + 1} + \frac{1}{n_5 + 1}$$
$$+ \frac{1}{n_6 + 1} + \frac{1}{n_7 + 1} + \frac{1}{n_8 + 1} + \frac{1}{n_9 + 1} + \frac{1}{n_{10} + 1}.$$

If the relationship in Equation (3.3) were not true and the summation were less than 1, as in the hypothetical case we first considered, then the punter could guarantee a win. The skill of a bookmaker is not just in fixing his odds to satisfy the golden rule — anyone with modest mathematical skills could do that. He must also properly assess the likelihood of each horse winning the race. If he fixed the odds on a horse at 10/1, when the horse actually has a one-in-five chance of winning, then astute professional gamblers would soon pick this up and take advantage of it.

3.3.   The Best Conditions for a Punter

We have seen that the bookmaker's golden rule, Equation (3.3), will ensure that no punter could place bets so as to guarantee to win on a particular race. If the quantity on the left-hand side were less than 1, then the punter could guarantee a win and if it is equal to 1, then the punter could place bets so as to guarantee getting exactly his stake money returned, but, of course, that would be a pointless exercise. Given that the left-hand side is greater than 1, then the larger its value, the larger the likely profit is that the bookmaker will make on the race. Logically, what is good for the bookmaker is bad for the punter, so we may make the proposition that, assuming that the relative chances of the horses winning are properly reflected in the odds offered, which is usually the case, then the smaller the summation, the less the odds are stacked *against* the punter. Conversely, the higher the summation, the better it is *for* the bookmaker. Let us consider the following two races with horses and odds as shown below. The value of $1/(n+1)$ is shown in brackets:

Race 1	Diana	2/1		(0.333)
	Dawn Lady	4/1		(0.200)
	Fairy Princess	6/1		(0.143)
	Olive Green	10/1		(0.091)
	Mayfly	10/1		(0.091)
	Dawn Chorus	15/1		(0.062)
	Missy	15/1		(0.062)
	Lovelorn	20/1		(0.048)
	Helen of Troy	25/1		(0.038)
	Piece of Cake	30/1		(0.032)
Race 2	Pompeii	3/2	($=1\frac{1}{2}/1$)	(0.400)
	Noble Lad	2/1		(0.333)
	Gorky Park	5/1		(0.167)
	Comet King	8/1		(0.111)
	Valiant	12/1		(0.077)
	Park Lane	16/1		(0.059)

The value of the summation for race 1 is 1.100 and for the second race it is 1.147. Other things being equal, and assuming that the relative chances of the horses winning are accurately reflected in the odds, the punter would have a better chance of winning by betting on his fancied horse in the first race.

One factor that works to the advantage of bookmakers is that a considerable amount of betting is done on an illogical basis, by whims or hunches. There are a number of classic races during the year, for example the Derby (in Epsom, UK), the St. Leger (in Doncaster, UK) and the Grand National, that attract large numbers of people, who normally never gamble on horses, to place bets either individually or through a sweepstake organized by a social club or at work. Most people placing bets on such occasions know nothing of the horses but may be attracted by the fact that they are trained locally or that their names are the same or similar to that of a friend or family member. For example, in 1948, Sheila's Cottage won the Grand National at odds of 50/1 and was backed by large numbers of punters lucky enough to have a wife, daughter or girlfriend called Sheila.

In contrast to the occasional or recreational punter, there are some professional gamblers who, by carefully assessing the odds, studying form and betting only when conditions are most favourable, manage to make a living from gambling. For them the big classic races have no particular attraction. One hundred pounds won on a minor race at a minor racetrack is the same to them as one hundred pounds won on a classic race, and if the classic race does not offer them a good opportunity, then they will simply ignore it.

It must always be remembered that bookmakers also make a living, and usually a good one, so for the average punter, gambling is best regarded as a source of entertainment and betting limited to what they can afford to put on.

Problem 3

3.1 The runners and odds offered by a bookmaker for three races at a particular track are as follows:

2:30 pm	Bungle	1/1 (Evens)
	Crazy Horse	4/1
	Memphis Lad	6/1
	Eagle	8/1
	Galileo	12/1
	Mountain Air	18/1
3:15 pm	Diablo	2/1
	Copper Boy	3/1
	Dangerous	3/1
	Boxer	4/1
	Kicker	6/1
	Zenith	8/1
3:45 pm	Turbocharge	2/1
	Tantrum	3/1
	Argonaut	4/1
	Firedance	8/1
	Dural	12/1

Do the odds for all these races conform to the bookmaker's golden rule? Which is the race that best favours the bookmaker's chance of winning?

Chapter 4

Making Choices and Selections

Refuse the evil, and choose the good (Isaiah 7:14)

4.1. Children Leaving a Room

We imagine that there are three children, Amelia, Barbara and Christine, playing in a room. The time comes for them to leave and they emerge from the room, one by one. In how many different orders can they leave the room? Using just their initial letters, we show the possible orders in Fig. 4.1.

You can easily check that there are no more than these six possibilities. If there had been another child in the room, Daniella, then the number of different orders of leaving would have been considerably greater — *viz.*:

ABCD	ABDC	ACBD	ACDB	ADBC	ADCB
BACD	BADC	BCAD	BCDA	BDAC	BDCA
CABD	CADB	CBAD	CBDA	CDAB	CDBA
DABC	DACB	DBAC	DBCA	DCAB	DCBA

While it is possible to generate these twenty-four orders of leaving in a systematic way, it gets more and more complicated as the number of children increases. Again, there is not much point in generating the detailed orders of leaving if all that is required is to know how many different orders there are.

For three children there were six orders of leaving and for four children there were twenty-four. Let us see how this could be derived

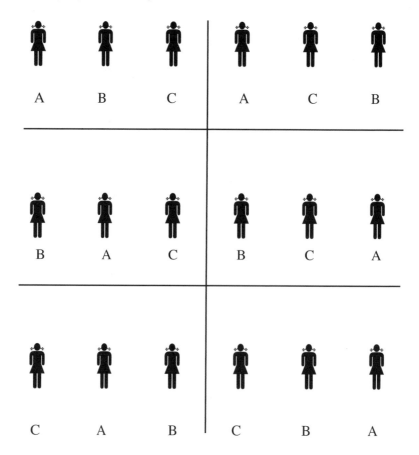

Fig. 4.1. The six orders in which three girls can leave a room.

without writing down each of the orders as has been done above. For three children there are three different possibilities for the first child to leave: A, B or C, corresponding to the different rows of Fig. 4.1. Once the first child has left, then there are two children remaining in the room and hence two possibilities for the exit of the second child, corresponding to the second letter for each of the orders. This leaves one child in the room, who finally leaves. Seen in this way, the number of possible patterns of departure is given by:

$$N_3 = 3 \times 2 \times 1 = 6. \tag{4.1}$$

If we repeat this pattern for four children, there are four possibilities for the first to leave, three for the second, two for the third and then the last one leaves. The number of possible patterns of departure for four children is given by:

$$N_4 = 4 \times 3 \times 2 \times 1 = 24. \tag{4.2}$$

This kind of approach gives the results that we found by listing the orders in detail and extending it to, say, seven children. Then the number of different orders of leaving is:

$$N_7 = 7 \times 6 \times 5 \times 4 \times 3 \times 2 \times 1 = 5{,}040. \tag{4.3}$$

It would be possible, but very complicated, to write down each individual order as was done for three children and four children.

Fully writing out these products of numbers for larger numbers of children is tedious and takes up a great deal of space, so as a shorthand, we write:

$$7 \times 6 \times 5 \times 4 \times 3 \times 2 \times 1 = 7! \tag{4.4}$$

where 7! is called *factorial* 7. For factorials of large numbers, say 50!, the conciseness of the notation is of great benefit.

Notice that we can find the number of orders, even when determining and listing the individual orders would not be practicable. For example, the number of orders in which a class of 30 pupils could leave a classroom is 30!, which is a number with 33 digits. A computer generating one billion (one thousand million) of these orders per second would take more than eight million billion years to complete the job!

4.2. Picking a Team

We consider a bridge club, consisting of twelve members, that has entered a national bridge tournament in which they will be represented by a team of four players. The members of the club are all of equal merit, so it is decided to pick the team by drawing lots, meaning that all possible selections of four players have the same probability of occurring. The question we ask is: "How many different compositions of teams is it possible to select from the members of the club?"

This problem can be considered in stages by picking members of the team one by one, much as we considered the problem of the children leaving the room. The first member of the team can be picked in twelve ways. For each of those twelve ways there are eleven ways of picking the next team member, so there are $12 \times 11 = 132$ ways of picking the first two members of the team. The next member can be picked in ten ways and the final member in nine ways, giving a total number of different ordered selections:

$$S = 12 \times 11 \times 10 \times 9 = 11{,}880. \qquad (4.5)$$

This will be the answer if the players have to be designated in some order because the tournament requires that players of a team must be labelled 1, 2, 3 and 4 to decide on how they play in pairs against opponents. However, if the team is just a group of four individuals, without regard to the order of their selection, then the value of S given by (4.5) is too great. Representing individuals by letters, the value of S comes about if selection ABCD is regarded as different from ACBD and all the other different ways of ordering the four letters. Within the 11,880 ordered selections there are groups of the same four individuals. To find the number of teams, without regard to order, one must divide S by the number of different ways of ordering four letters. This is precisely the problem of the number of different orders in which four girls can leave a room; the first letter can be chosen in four ways, then the second in three ways, the third in two ways, and finally, the fourth in one way, so the number of ways of ordering four letters is:

$$4 \times 3 \times 2 \times 1 = 4!$$

From this we find the number of teams, *without regard to order of selection*, as:

$$T = \frac{S}{4!} = \frac{12 \times 11 \times 10 \times 9}{4!} = 495. \tag{4.6}$$

It is possible to express this in a more succinct way. We note that:

$$12 \times 11 \times 10 \times 9 = \frac{12 \times 11 \times 10 \times 9 \times 8 \times 7 \times 6 \times 5 \times 4 \times 3 \times 2 \times 1}{8 \times 7 \times 6 \times 5 \times 4 \times 3 \times 2 \times 1}$$

$$= \frac{12!}{8!},$$

giving:

$$S = \frac{12!}{8!} \tag{4.7}$$

and then

$$T = \frac{S}{4!} = \frac{12!}{8!4!}.$$

We can generalize from this example. Let us suppose that there are n objects, distinguishable from each other — for example, numbered balls contained in a bag. If we take r of them out of the bag, one at a time, and take the order of selection into account, then the number of different selections is:

$$S = \frac{n!}{(n-r)!}. \tag{4.8}$$

By making $n = 12$ and $r = 4$ we get (4.7): the number of ways of selecting an ordered team of four people from twelve. However, if the order is not taken into account, corresponding to putting your hand in the bag and taking a handful of r balls, then the number of different selections is:

$$T = \frac{n!}{(n-r)!r!} \tag{4.9}$$

that, with $n = 12$ and $r = 4$ is the number of different teams of four people one can select from twelve without order being taken into account. The expression on the right-hand side of (4.9) is important enough to have its own symbol, so we can write *the number of combinations of r objects taken from n* as:

$$^nC_r = \frac{n!}{(n-r)!r!}. \tag{4.10}$$

It is obvious that the number of combinations of n objects that can be taken from n is one: you take all the objects and that is the only combination. Inserting $r = n$ in (4.10) gives:

$$^nC_n = \frac{n!}{0!n!} = 1,$$

from which we conclude the interesting result that:

$$0! = 1, \tag{4.11}$$

which is not at all obvious.

4.3.   Choosing an Email Username

The most common number of initials designating an individual in Western societies is three, corresponding to a family name plus two given names. Thus Gareth Llewelyn Jones would be recognized by his initials "GLJ", with which he would sanction alterations on his cheques, or which he would put at the bottom of notes to colleagues at work. However, when Gareth signs up to an internet provider and asks for the username "GJL" he is told that it is already in use. So he adds the digit "1" to his username and finds that it is still unavailable. He then tries the digit "2" with the same result, as happens for "3", "4" and "5". Why is this?

Let us assume that the internet provider has one million customers wishing to use a set of three initials as a username. We first ask ourselves how many different sets of three initials we could have on the assumption that all letters of the alphabet are equally likely, which is not actually true, since letters such as "Q", "X" and "Z" rarely occur in English. Each initial has 26 possibilities so the number of possible combinations is:

$$26 \times 26 \times 26 = 17{,}576.$$

Hence the expected number of times the combination "GLJ" would appear in one million customers is:

$$N = \frac{1{,}000{,}000}{17{,}576} = 57 \qquad (4.12)$$

to the nearest whole number. Actually, given the fact that "G", "L" and "J" are fairly common initials, the number of people in the provider's customer list sharing Gareth's combination may well be closer to 100. If Gareth were to try something like "GLJ123", then he would be likely to find a unique username.

Even a four-letter username is likely to need some appended digit to make it unique since $26^4 = 456{,}976$, less than one million.

4.4.   The UK National Lottery

A single entry in the UK National Lottery consists of picking six different numbers in the range 1 to 49. If these match the numbers on the balls picked at random out of a drum, then the first prize is won. A seventh ball is also selected from the drum, the *bonus ball*; if you have the bonus ball number, plus any five of the six main winning numbers, then you win the second prize. There are also smaller prizes for getting five correct numbers without the bonus ball number, and also for four or three correct numbers. Figure 4.2 shows some examples for winning each kind of prize.

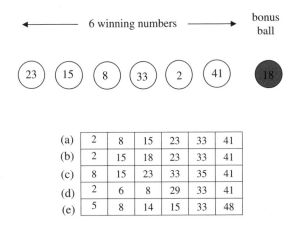

Fig. 4.2. The selected numbered balls and some winning lines: (a) first prize: 6 correct (b) second prize: 5 correct plus bonus ball (c) third prize: 5 correct (d) fourth prize: 4 correct (e) fifth prize: 3 correct.

What are the chances of winning the first prize with a single entry? All combinations of six different numbers are equally likely and the number of combinations of six that can be selected from forty-nine is obtained from (4.10) with $n = 49$ and $r = 6$, giving:

$$^{49}C_6 = \frac{49!}{43!6!} = \frac{49 \times 48 \times 47 \times 46 \times 45 \times 44}{6 \times 5 \times 4 \times 3 \times 2 \times 1} = 13{,}983{,}816. \quad (4.13)$$

Since there is only one correct combination, the probability of winning the first prize is:

$$P_{first} = \frac{1}{13{,}983{,}816}. \quad (4.14)$$

It is probably sensible to await the draw before ordering the Rolls Royce!

To win a second prize you must have five of the six winning numbers, plus the bonus ball number. The number of different ways of having five of the six winning numbers is six. For the example of the six winning numbers given in Fig. 4.2 these are:

$$
\begin{array}{ccccc}
2 & 8 & 15 & 23 & 33 \\
2 & 8 & 15 & 23 & 41 \\
2 & 8 & 15 & 33 & 41 \\
2 & 8 & 23 & 33 & 41 \\
2 & 15 & 23 & 33 & 41 \\
8 & 15 & 23 & 33 & 41
\end{array}
$$

Since there are six different ways of having five of the six winning numbers, so there are six different combinations: five winning numbers + the bonus ball number that can win the second prize. Thus the probability of winning this way is the ratio of the number of ways of having a winning combination, six, divided by the total number of different combinations, that is:

$$
P_{second} = \frac{6}{13,983,816} = \frac{1}{2,330,636}. \tag{4.15}
$$

We have seen that there are six different ways of having five winning numbers and the number of ways of having a sixth number which is neither a winning number nor the bonus ball number is:

49 − number of winning numbers − bonus ball number

$$
= 49 - 6 - 1 = 42.
$$

Thus the number of combinations for winning with five correct numbers is $6 \times 42 = 252$ and hence the chance of winning this way is:

$$
P_{third} = \frac{252}{13,983,816} = \frac{1}{55,491.33}. \tag{4.16}
$$

The number of ways of winning a prize with four correct numbers is the number of combinations of having four out of six winning numbers with two out of forty-three non-winning numbers. The number of ways of having four out of the six winning numbers is:

$$^6C_4 = \frac{6!}{4!2!} = 15$$

and the number of ways of having two out of forty three non-winning numbers is:

$$^{43}C_2 = \frac{43!}{2!41!} = 903.$$

Hence the number of combinations with just four correct numbers is:

$$^6C_4 \times {}^{43}C_2 = 15 \times 903 = 13{,}545,$$

giving the probability of winning as:

$$P_{fourth} = \frac{13{,}545}{13{,}983{,}816} = \frac{1}{1{,}032.4}. \qquad (4.17)$$

Finally, to have just three winning numbers you need three out of six winning numbers together with three out of forty-three non-winning numbers. The number of ways of doing this is:

$$^6C_3 \times {}^{43}C_3 = \frac{6!}{3!3!} \times \frac{43!}{40!3!} = 246{,}820,$$

so the chance of winning this way is:

$$P_{fifth} = \frac{246{,}820}{13{,}983{,}816} = \frac{1}{56.66}. \qquad (4.18)$$

The money staked on the National Lottery is used for three purposes. Some of it is returned to the prize winners, some provides a profit for the company that organizes and runs the lottery and the remainder goes to "good causes". There are ways to contribute to good causes while having a "flutter" that are more cash efficient, but there is no doubt that many enjoy the anticipation of the periodic drawing of the winning balls one by one from a drum, performed on television with flashing lights and theatrical music, and have the dream that one day the big prize will be theirs.

Problems 4

4.1 Five children are in a room and leave in single file. In how many different orders can the children leave?

4.2 Ten members of a darts club are better than all the other members and are all equally proficient. It is required to select a team of four from these ten to represent the club in a tournament and the selection is done by drawing lots. In how many different ways can the team be chosen?

4.3 A bag contains six balls all of different colours. Three balls are taken out of the bag, one at a time. What is the probability that the balls selected will be red, green and blue in that order? The three selected balls are returned to the bag and then a handful of three balls is taken out. What is the probability that the three balls are red, green and blue?

4.4 A town decides to run a local lottery based on the same model as the National Lottery. There are only twenty numbers from which to pick, 1 to 20, and only four numbers are picked. Four winning numbers are selected from a drum full of numbered balls plus a bonus ball. The top prize is for four correct numbers, the second prize is for three correct numbers plus the bonus ball and the third prize is for three correct numbers. What are the probabilities of winning each of these prizes with a single entry in the lottery?

Chapter 5

Non-Intuitive Examples of Probability

And things are not what they seem (Henry Wadsworth Longfellow, 1878, "A Psalm of Life")

We have already mentioned that it is intuitively obvious that the probability of obtaining a head when flipping an unbiased coin is 0.5 and that of getting a 6 when throwing a die is $\frac{1}{6}$, but sometimes intuition can lead you astray. Here we shall deal with three situations where, to most people, the probabilities of particular outcomes are not at all obvious.

5.1. The Birthday Problem

Ignoring the complication of leap days, there are 365 days on which a birthday can occur. A moment of thought tells us that if there are 366 people in a room, then it is *certain* (probability = 1) that at least one pair of them must share a birthday. If the number of people were reduced to 365 or 364, then while it is not certain that at least one pair share a birthday, our intuition tells us that the probability of this must be very close to 1. Indeed, we should be astonished if no two people shared a birthday with such numbers. If the number of people were reduced further then, clearly, the probability would be reduced but would still remain high down to 200 people or so.

Let us now start at the other end with two people in a room. The probability that they have the same birthday is low and is easy to calculate. One way of thinking about it is to consider that one of them tells you their birthday. Then you ask the other one for their birthday. There are 365 possible answers they can give but only one of them will match the birthday given by the first person. Hence the probability of the birthdays being the same is: $\frac{1}{365}$.

The other possible way to look at this problem is to first ask the question: "In how many possible ways can two people have birthdays without any restriction?" Since there are 365 possibilities for each of them, then the answer to this is 365×365: every date in the calendar for one, combined with every date in the calendar for the other. Now we are going to ask a question that is not, perhaps, the obvious one to ask: "In how many ways can the two people have different birthdays?" Well, the first one has 365 possibilities but once their birthday is fixed then, for the birthdays to be different, the second has only 364 possibilities, so the answer is 365×364. Now we can find the *probability* that they have *different* birthdays as:

$$p_{diff} = \frac{\text{Number of ways of having different birthdays}}{\text{Total number of ways of having birthdays}}$$

$$= \frac{365 \times 364}{365 \times 365} = \frac{364}{365}. \tag{5.1}$$

Now, either the two individuals have the same birthday, or they do not and these are the only possibilities and they are mutually exclusive so that:

$$p_{diff} + p_{same} = 1$$

or

$$p_{same} = 1 - p_{diff} = 1 - \frac{364}{365} = \frac{1}{365}, \tag{5.2}$$

the answer obtained previously. The reason we have introduced this apparently convoluted way of getting the answer is because it better lends itself to dealing with the problem of when there are three or more people in the room.

Now consider three people in a room. The number of ways they can have birthdays, without any restriction, is $365 \times 365 \times 365$. The number of ways they can have *different* birthdays is found by noting that the first person can have a birthday in 365 ways. For the second to be different from the first there are 364 ways, and for the third to be different from the other two there are 363 ways. Hence the total number of ways they can have different birthdays is: $365 \times 364 \times 363$. Hence the probability that they have different birthdays is thus:

$$3p_{diff} = \frac{\text{number of ways birthdays can be different}}{\text{number of ways they can have birthdays}}$$

$$= \frac{365 \times 364 \times 363}{365 \times 365 \times 365} = 0.9917958. \tag{5.3}$$

If the birthdays are not all different, then at least two of them must be the same and the probability of this is:

$$3p_{not\ diff} = 1 - 3p_{diff} = 0.0082042. \tag{5.4}$$

The prefix 3 before the p is to indicate that the answer is for three people.

We should now be able to see by extending the principle that the probability that at least two people out of five people in a room would share a birthday is:

$$5p_{not\ diff} = 1 - \frac{365 \times 364 \times 363 \times 362 \times 361}{365 \times 365 \times 365 \times 365 \times 365} = 0.027136. \tag{5.5}$$

The advantage of calculating the probability in the way that we do (by finding the probability that all birthdays are different and then subtracting from 1) is now evident. Suppose that we tried directly to find the probability that two or more people in a set of five shared a birthday. We would have to consider one pair the same and the others all different; three with the same birthday with the other two different; two pairs with the same birthday, and many other combinations. All birthdays being *different* is a single outcome, the probability of which is comparatively easy to find.

Now we ask the question that is really the basis of this section: "How many people must there be in a room for there to be a probability greater than 0.5, so that at least two of them would share a birthday?" This can be done with a calculator. We work out the following product term by term, until the answer we get is *less than* 0.5:

$$\frac{365}{365} \times \frac{364}{365} \times \frac{363}{365} \times \frac{362}{365} \times \frac{361}{365} \times \frac{360}{365} \times \frac{359}{365} \times \cdots\cdots\cdots\cdots\cdots$$

The value of n, for which the answer first falls below 0.5, is when the probability of all birthdays being *different* is less than 0.5 and hence when there is a probability greater than 0.5 that two or more people have the *same* birthday. This is the number of people that fulfils our requirement. What is your intuitive assessment of n at this point?

In Table 5.1 we give the probability of having at least two people with the same birthday with increasing n. We see that the probability just exceeds 0.5 when 23 individuals are present: a result that most people find surprisingly low. A visual impression of what this result gives is illustrated in Fig. 5.1.

Table 5.1. The probability that two or more people have the same birthday with n people present.

n	Probability	n	Probability	n	Probability
2	0.002740	3	0.008204	4	0.016356
5	0.027136	6	0.040462	7	0.056236
8	0.074335	9	0.094624	10	0.116948
11	0.141141	12	0.167025	13	0.194410
14	0.223103	15	0.252901	16	0.283604
17	0.315008	18	0.346911	19	0.379119
20	0.411438	21	0.443688	22	0.475695
23	0.507297	24	0.538344	25	0.568700

Fig. 5.1. There is a greater than 50% chance that two of these people will share a birthday!

For someone teaching a class of 30 students, the probability that two or more will share a birthday is 0.706316; for 40 students it's 0.891232, and for 50 students 0.970374.

5.2.   Crown and Anchor

Crown and Anchor is a dice game that was once popular with sailors in the British Navy. It was played with three dice, each of which had on its six faces a crown, an anchor and the symbols for the card suits: club, diamond, heart and spade (Fig. 5.2).

However, the essence of the game is preserved if the dice are normal ones with the numbers 1 to 6 on their faces, and we shall consider them so. There are two players: the banker who runs the game and the thrower who rolls the dice. We start by considering a game that is *not* Crown and Anchor.

The banker first justifies the fairness of the game to the intended non-mathematical victim, who is the thrower-to-be. He explains that since the probability of getting a 6 with one die is $\frac{1}{6}$, then the probability of getting a 6 when throwing three dice is $\frac{3}{6} = \frac{1}{2}$. Of course,

Fig. 5.2.   A set of three Crown and Anchor dice.

that is complete nonsense, since adding probabilities is only justified when the outcomes are mutually exclusive. However, the outcomes from the three dice are actually *independent*, since getting a 6 on one die has no influence over what happens with the other two dice. However, our thrower-to-be has not read this book, so accepts the argument and throws on the basis that if he gets a 6, the banker pays him £1, and if he fails to get a 6, then he will pay the banker £1.

The thrower loses heavily and it gradually dawns on him that something is wrong. Sometimes he throws two 6s and once in a while three 6s but he still gets paid just £1. He remonstrates with the banker and asks to be paid £2 if he gets two 6s, and £3 if he gets three 6s. With a show of reluctance, the banker agrees and they continue to play what is now truly the game of Crown and Anchor. The thrower still loses, albeit a little slower than previously, but he concludes that since the game is now fair, and being played under the rules he demanded, he must just be having an unlucky day. Is he really unlucky or has he been duped? Let us analyse the game.

The first thing we must determine is what outcomes are possible from throwing three dice. Representing not-a-6 by the symbol 0, these possibilities for the three dice are:

$$000 \quad 600 \quad 060 \quad 006 \quad 660 \quad 606 \quad 066 \quad 666$$

The probability of getting a 6 with any die is $\frac{1}{6}$ and the probability of getting not-a-6 is $\frac{5}{6}$. Since the outcomes from the three dice are independent, the probabilities of the outcomes listed above are given in the second column of Table 5.2 by multiplying probabilities.

It will be found that the sum of the probabilities in the second column is 1, as indeed they must be, since the mutually exclusive outcomes listed are the totality of possible outcomes. When the first game, in which the thrower is not rewarded for multiple 6s, is played 216 times with a unit stake, then on 125 occasions the banker wins, and on the remaining 91 occasions the thrower wins. This gives a net loss of 34 stake units. This is shown in the third column of the table. However, when the game is played according to the rules of

Table 5.2. The probabilities for various outcomes of throwing three dice and the expected profits and losses for the thrower for each outcome, playing the first game 216 times and then playing Crown and Anchor 216 times.

Outcome	Probability	Profit (loss) on first game	Profit (loss) on C&A
0 0 0	$\dfrac{5}{6} \times \dfrac{5}{6} \times \dfrac{5}{6} = \dfrac{125}{216}$	(-125)	(-125)
6 0 0	$\dfrac{1}{6} \times \dfrac{5}{6} \times \dfrac{5}{6} = \dfrac{25}{216}$	25	25
0 6 0	$\dfrac{5}{6} \times \dfrac{1}{6} \times \dfrac{5}{6} = \dfrac{25}{216}$	25	25
0 0 6	$\dfrac{5}{6} \times \dfrac{5}{6} \times \dfrac{1}{6} = \dfrac{25}{216}$	25	25
6 6 0	$\dfrac{1}{6} \times \dfrac{1}{6} \times \dfrac{5}{6} = \dfrac{5}{216}$	5	10
6 0 6	$\dfrac{1}{6} \times \dfrac{5}{6} \times \dfrac{1}{6} = \dfrac{5}{216}$	5	10
0 6 6	$\dfrac{5}{6} \times \dfrac{1}{6} \times \dfrac{1}{6} = \dfrac{5}{216}$	5	10
6 6 6	$\dfrac{1}{6} \times \dfrac{1}{6} \times \dfrac{1}{6} = \dfrac{1}{216}$	1	3
	Net loss	34	17

Crown and Anchor, the thrower receives two stake units when they throw two 6s and three stake units when they throw three 6s but, as seen in the final column of Table 5.2, they still have a loss of 17 stake units. The game seemed fair — why is the thrower still losing?

If three dice are thrown 216 times, then altogether $3 \times 216 = 648$ faces are seen, one sixth of which, i.e., 108, are expected to be 6s. Now let us suppose that when 6s do appear they only appear singly, so that two 6s and three 6s never happen. In this case, the thrower will win 108 times — each time a 6 appears — and the banker will win 108 times — each time a 6 does not appear — so there will be no profit or loss on either side. Now imagine that two of the expected 108 6s do appear together and the other 106 6s appear singly. The thrower still wins 108 units, 1 unit for each appearance of a 6, but they have won this on 107 throws, so now there are 109 occasions when there is no 6. The banker is now 1 unit in profit. This is the secret of the game: every time that 6s appear twice or three times, no doubt welcomed jubilantly by the thrower, the conditions are created for extra gains by the banker. All is not what it seems in Crown and Anchor.

In the early years of the twentieth century, and perhaps earlier, bankers would set up street games of Crown and Anchor in London and other big cities. The variation from what has been described is that the thrower could choose the symbol on which they were staking, for example, the anchor, but that made no difference to the probabilities involved. This was an illegal activity, since the only form of gambling that was legal at that time was betting on horses or greyhounds at registered tracks and in registered casinos and clubs. Crown and Anchor is a game so heavily biased against the thrower that it is tantamount to robbing the innocent and so should be banned even under present, more liberal legislation.

5.3. To Switch or Not to Switch, That is the Question

Television game shows in which contestants select boxes containing various rewards are very popular. Pick the right box and a fortune is yours, pick the wrong box and the prize is derisory or even something unpleasant. The games never involve the single simple act of selecting a box but are usually accompanied by a set of intermediate actions or decisions before the box is finally opened. We now describe one such game.

The contestant is offered three boxes, which we shall indicate as A, B and C. In one of the boxes is £1,000, while the other two boxes each contains a sausage. The contestant picks box A but does not open it. The game show host then opens one of the other two boxes, which is seen to contain a sausage. Now the host archly asks the contestant whether he would like to stick with his original box or switch to the other unopened box. The contestant is suspicious of this offer. He concludes that since there are two boxes unopened, and one of them contains the money, the probability of the money being in each unopened box is 0.5, so he might just as well stick with his original choice. The argument seems impeccable but, in fact, he has made the wrong choice. His chance of winning is actually doubled by deciding to change to the other unopened box. A surprising result, but let us see how we get to it.

The critical factor in this game is that the host knows where the money is and will always open a box containing a sausage. Now, let us assume that the contestant initially chooses box A. The possible content of each of the boxes in terms of money and sausages is as shown in Fig. 5.3.

First we suppose that the contestant decides not to switch boxes. If the arrangement is the top one in Figure 5.3, then he will win the money. However, if the arrangement is one of the other two, then he will find a sausage in his box. His chance of winning is therefore 1 in 3, i.e., 1/3. Next, we assume that he always decide to switch boxes. If the arrangement is the top one, then whichever sausage the host decides to display, our contestant will have picked the other sausage and so by switching he has lost out. But for the middle arrangement, the host will have opened box C, the switch will give the contestant box B and he will win the money. Likewise,

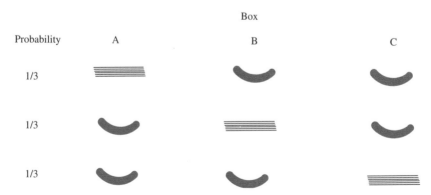

Fig. 5.3. The three possible arrangements of money and sausages, each with probability 1/3.

for the bottom arrangement, the host will have opened box B, our contestant will have switched to box C and again he wins the money. For two out of the three possible arrangements he wins, so his probability of winning is 2/3, double what it was if he did not switch.

Since all the boxes are effectively equivalent, it is arbitrary which is designated as A, B and C: the result we have found is quite independent of which box the contestant originally chose. He always has a better chance of winning the money if he decides to switch boxes. A surprising result, you will agree.

Problems 5

5.1 On the planet Arret there are 100 days in a year.

   (i) What is the probability that two Arretians, chosen at random, will have different birthdays?

   (ii) What is the probability that four Arretians, chosen at random, will have different birthdays?

   (iii) How many Arretians must there be in a room for there to be a probability greater than 0.5 that at least two of them share a birthday?

5.2 A game of Crown and Anchor is played with two dice. In the terminology of Table 5.2 the possible outcomes are: 0 0, 0 6, 6 0 and 6 6.

   Construct a table similar to Table 5.2, showing what the expected profit is for the banker from 36 games, for both the first game and the two-dice Crown-and-Anchor game.

Chapter 6

Probability and Health

If you believe the doctors, nothing is wholesome (Lord Salisbury, quoted in a letter by Lady Gwendolen Cecil to Lord Lytton, 15th June 1877)

6.1. Finding the Best Treatment

Medicine is a field of activity in which the type of probability of interest is empirical probability, as described in Problem 1.4 and in Section 3.1. Bodies, such as national medical councils and the World Health Organization, collect and collate information from different sources and so derive empirical fatality rates for various diseases and success rates for different clinical treatments and surgical procedures. Of course, in each individual case where diagnosis and treatment are required, other factors come into play, such as the training, skill and experience of the doctor or surgeon and also the characteristics of the patient, for example, age and robustness. In view of all these factors there is some element in diagnosis and treatment that is akin to what was called *judgemental probability* in Section 3.1. Trying to introduce some certainty in diagnosis, various computer-based diagnostic systems have been devised. When the computer diagnoses for four of these systems were tested against the diagnoses produced by a group of highly experienced and competent doctors for particular sets of symptoms, the programmes gave a correct diagnosis in one-half to three-quarters of cases. Where there were several possible conditions suggested by the symptoms, then the programmes gave less than one-half of the possibilities suggested by the experts. By contrast, on average, the programmes suggested two additional diagnoses per case that the experts had not themselves proposed, but which they thought were relevant.

Despite all the uncertainties of diagnosis, more often than not when we visit a doctor with a medical condition the doctor will recognize the problem and know the treatment that will best deal with it. But, as we have already indicated, that is not always so. Sometimes symptoms may not point uniquely to a particular medical condition but may indicate one or other of a finite number of possible conditions with different probabilities of being present (Fig. 6.1).

We shall consider a hypothetical case of an uncertain diagnosis indicating three possible conditions, A, B and C with probabilities:

A   0.70   B   0.20   C   0.10

These probabilities will be based on a considerable body of information gathered by doctors and medical bodies and published in

Fig. 6.1.   "I'm not really sure what's wrong with you but I know the best way to treat it."

research journals. There is a range of drugs available for treating these conditions and, since the symptoms are similar, the drugs will probably all have some efficacy for each of the possible conditions. We shall assume that there are three drugs available, a, b and c, which have the following observed probabilities of dealing successfully with each of the conditions:

Drug a   A   0.6   B   1.0   C   0.4
Drug b   A   0.65   B   0.5   C   0.9
Drug c   A   0.75   B   0.2   C   0.5

From the probabilities of the conditions, it is clear that A is most likely to be present, and from the effectiveness of the drugs, it is clear that c is the best for treating condition A. So should we conclude that, from the point of view of the patient, the best treatment is to administer drug c? The answer is no: it is actually the worst decision to make!

To see what the doctor should prescribe, let us make the absurd, but numerically helpful, assumption that she has 1,000 patients with this uncertain condition and then work out the likely outcome of the three drug treatments. Of the 1,000 patients we would expect the following numbers with the three conditions:

A   700   B   200   C   100

If they use drug a then:

of the 700 patients with condition A, a number $700 \times 0.6 = 420$ will recover,

of the 200 patients with condition B, a number $200 \times 1.0 = 200$ will recover,

and

of the 100 patients with condition C, a number $100 \times 0.4 = 40$ will recover.

Hence the total number of recovered patients would be: $420 + 200 + 40 = 660$. We now repeat this exercise for the other two drugs. The outcomes are:

For drug b the number recovering is: $700 \times 0.65 + 200 \times 0.5 + 100 \times 0.9 = 645$.

For drug c the number recovering is: $700 \times 0.75 + 200 \times 0.2 + 100 \times 0.5 = 615$.

The best drug to use, drug a, turns out to be the one that is least effective for the most likely condition, A, but it is far more effective for the other two conditions.

What we learn from this is that in cases where there are many possible mutually exclusive situations (i.e., conditions A, B and C) with different probabilities, and different possible reactions to those situations (i.e., applying drugs a, b and c), with different probabilities of having a successful outcome, it is necessary to consider the outcomes of all possible combinations of situations and reactions to optimize the chances of success. Here we have met another situation, like those described in Chapter 5, where the answer obtained from probability theory may be at variance with one's intuition.

6.2.   Testing Drugs

In non-Western societies, notably in China and India, branches of medicine have evolved over many centuries that involve the use of natural products, usually of vegetable, but sometimes of animal, origin. Many of these treatments are very effective and provide the basis for the synthetic drugs used in Western medicine. However, there are many advantages in the use of synthetic drugs rather than natural products. First, the dosage can be more accurately controlled; the concentration of the active ingredient in a natural product, say, in the root of a plant, may depend on the weather conditions during the growing season. Again, by extracting and analysing the structure of the active ingredient in a plant, a drug company may be able to produce derivatives that are either more effective than the natural product or have less severe side effects, or both. Another modern aspect of the pharmaceutical industry is that of drug design. The beneficial effect of many drugs is obtained by their action on particular proteins in the patient and drugs can be designed so that they bind themselves in specific ways to the protein and so either inhibit or increase its activity.

Many drugs are tested on animals, where the assumption is that the reaction of the animal, say, a guinea pig or a mouse, to the drug will be similar to that of a human patient. Often it is, but sometimes it is not, so at the end of the day the only certain way to determine the effect on human patients is to test it on human patients. Occasionally, and fortunately rarely, such tests can go badly wrong, as happened with the test of a drug in the UK in March 2006. Tests on mice and primates had shown beneficial medical results without side effects. However, six men given the drug in a trial all became seriously ill, with failure of various organs and permanent damage to health. While this was a setback for testing drugs on humans, it is certain that such tests will continue, although with much more stringent safeguards than hitherto.[a]

It is important that when such trials are carried out, a proper assessment should be made of the effectiveness of the new procedure. The theory behind these assessments is fairly difficult but we shall describe how the application of a few simple rules enables a decision to be made regarding the significance of the result of a clinical trial.

We will consider the hypothetical trial of a drug that is administered to 10 patients suffering from a particular disease. Of the 10 patients, 7 of them recover but 3 get no benefit from the treatment. Another group of 20 patients with the same disease are given a placebo, a harmless and inactive substance that in appearance

---

[a]Working Party on Statistical Issues in First-in-Man Studies, 2007, "Statistical Issues in First-in-Man Studies". *Journal of the Royal Statistical Society* **170A**, 517–579.

resembles the drug. Of these patients, 5 recover but 15 do not. The patients themselves do not know whether they received the drug or the placebo, so we can rule out any psychologically-induced physio-logical effects. Is there any evidence that the new drug is effective? To decide on this we are going to describe the steps in what statisticians call the χ^2-test (χ is a Greek letter written as "chi" but pronounced as "kye", so this is the chi-squared test).

Step 1   Represent the results of the trial in a table O (for observation) as shown below.

	Recover	Don't recover	
Drug taken	7	3	10
Placebo taken	5	15	20
	12	18	

Table O

    The totals given at the right-hand side of the rows are the total numbers of patients given the drug and given the placebo. The totals given at the bottom of the columns are the total numbers of recovered and non-recovered patients, respec-tively.

Step 2   We make a hypothesis (called the *null hypothesis*) that the drug is *ineffective*. Overall, 12 out of 30 patients, i.e., a frac-tion 0.4, recovered and *if the null hypothesis is valid* then this is the proportion of patients that recover spontaneously with-out any treatment. Based on the null hypothesis, we now produce a Table E (for expected), giving the expected num-bers of recovered and non-recovered patients for those given the drug (now assumed ineffective) and those not given the drug. Thus, of the 10 patients given the drug, on the basis of the overall proportion of recoveries, we would expect 4 to recover and 6 not to recover. Similarly, for the number of patients given the placebo, the expected number recovering and not recovering would be 8 and 12, respectively. Notice that when the expected numbers are entered in the table, the sums of rows and columns are the same as they were previously:

	Recover	Don't recover	
Drug taken	4	6	10
Placebo taken	8	12	20
	12	18	

Table E

Step 3 Now, even if the drug is completely ineffective, the O and E tables would probably be different just because of random fluctuations. As an example, if you flipped a coin 100 times it is unlikely that there would be *exactly* 50 heads and 50 tails. However, if you obtained 90 heads and 10 tails, you might suspect that your null hypothesis, that the coin was unbiased, was almost certainly untrue. Now we see how to test the differences between the O and E tables to see whether they are so different that we might suspect that our null hypothesis was untrue. To do this we calculate the value of:

$$\chi^2 = \sum_{i=1}^{4} \frac{(O_i - E_i)^2}{E_i}, \tag{6.1}$$

where $i = 1$ to 4 for the four entries in the table and O_i and E_i are the entries in the O and E tables, respectively. Remembering the explanation of $\sum_{i=1}^{4}$ given in relation to Equation (3.3),

Table 6.1. Probabilities associated with the χ^2 distribution for one degree of freedom.

Probability	0.20	0.10	0.05	0.025	0.02	0.01	0.005	0.001
χ^2		1.642	2.706	3.841	5.024	5.412	6.635	7.879	10.827

this value is:

$$\chi^2 = \frac{(7-4)^2}{4} + \frac{(3-6)^2}{6} + \frac{(5-8)^2}{8} + \frac{(15-12)^2}{12} = 5.625.$$

(6.2)

Step 4 The value of χ^2 is compared with the numbers in a standard table, part of which is shown in Table 6.1.

Let us imagine that our value of χ^2 had been 3.841. The table would tell us that if the null hypothesis is true and the drug is ineffective, then the probability of getting this value of χ^2, *or some greater value*, is 0.05, which is 1 in 20. What significance you place on that value is a subjective choice that depends on the kind of test being made. If the test related to some commercial decision, for example, testing consumer reaction to some new form of packaging, then possibly 0.05 would be regarded as sufficiently low to indicate that the null hypothesis, i.e., that the new packaging was ineffective, was probably untrue. In medicine the decision barrier is usually much more stringent and if the medical condition is a matter of life and death, then it would be extremely stringent. The value of χ^2 found in the example we have used indicates a probability just under 0.02 that the result could be obtained with the null hypothesis but this may not be small enough to pronounce that the treatment is effective and so sanction its use. It is the usual practice to decide on the level of significance that would lead to acceptance *before* doing the test. If that is not done, then there is always the danger that, with the result of the test already being known, a significance level will be set so as to achieve a desired outcome.

Let us suppose that a significance level of 0.01 had previously been set as a criterion for judging whether or not the new drug should be adopted for general use. This means that the value of χ^2 must be such that there is a probability less than 0.01 that the result of the test, or something even further from expectation, could have been obtained just by chance. The test described above would not have met the conditions for adoption. However, now let us suppose that the test had been carried out on a larger scale and that the observed and expected tables had been as shown below:

	Recover	Don't recover	
Drug taken	14	6	20
Placebo taken	10	30	40
	24	36	

Table O

	Recover	Don't recover	
Drug taken	8	12	40
Placebo taken	16	24	20
	24	36	

Table E

These tables are just the ones we considered previously scaled up by a factor of two and now we have:

$$\chi^2 = \frac{(14-8)^2}{8} + \frac{(6-12)^2}{12} + \frac{(10-16)^2}{16} + \frac{(30-24)^2}{24} = 11.25. \quad (6.3)$$

From Table 6.1 it is clear that there is less than 1 chance in 1,000 that the result of the test, or some more extreme result, could come about by chance, if the null hypothesis were true. The test satisfies the criterion for adoption of the drug and its efficacy is confirmed beyond all reasonable doubt. This shows the importance of scale in carrying out significance tests.

The χ^2 test can be applied in many different contexts and we have just given one example here. In the title of Table 6.1 there is a reference to "one degree of freedom". The concept of "degrees of freedom" is very important in statistics and is a measure of how many independent quantities there are in the system of interest. We noticed that although the O and E tables had different entries, the totals of both rows and columns are the same. Thus if we were told that 20 people had been given the drug and 40 given the placebo, and that 24 patients had recovered but that 36 had not, then that would not enable us to construct a table. Both the O and E tables satisfy the information given and many other tables can be constructed that do so. In this context one degree of freedom means that, *given the totals of rows and columns, once a single entry is made in the table, then all the other entries follow.*

We shall meet the concept of degrees of freedom again; there are other applications of the χ^2 test, which will be described in Chapter 8, that involve many degrees of freedom. While it is a difficult concept to understand fully, fortunately, in the contexts in which it occurs, it is usually simple to find out how many degrees of freedom there are.

Problems 6

6.1 The symptoms for a patient indicate one of two conditions: A with a probability of 0.65 or B with a probability of 0.35. Two drugs are available, both of which have partial success in dealing with both conditions. Their effectiveness is summarized in the following table which gives the success rate in giving a cure:

Condition	A	B
Drug a	0.6	0.4
Drug b	0.3	0.9

To maximize the probability of obtaining a cure, which drug should the doctor prescribe?

6.2 A chain store decides to test the effectiveness of a new form of packaging for one of its products. In two of its stores, in similar environments and with similar sales profiles, it arranges similar displays of the product. Of the first 100 customers that enter store A, 8 buy the product with the old packaging. Of the first 100 customers that enter store B, 20 buy the product with the new packaging.

Calculate the value of χ^2. The null hypothesis is that the new packaging is ineffective. If the probability that the value of χ^2, or some larger value, could be obtained with the null hypothesis just by chance, is greater than 0.05, then the store will not bother to change the packaging. Do they change it or not?

Chapter 7

Combining Probabilities: The Craps Game Revealed

Un coup de dés jamais n'abolira le hazard (A throw of a dice will never eliminate chance) (Stéphane Mallarmé, 1897, *Cosmopolis: An International Review*)

7.1. A Simple Probability Machine

We consider a simple probability machine, where a ball drops out of a hole and falls along a set of channels arranged in a pattern, as shown in (Fig. 7.1).

Whenever the ball reaches a junction point, shown by a small circle, it has an equal probability of taking the left-hand or right-hand path to the next line of junctions. At the bottom, indicated by the larger circles, is a set of cups, into one of which the ball falls. To start a ball on its journey a 10 p coin must be inserted in the machine. If the ball enters the centre cup, the money is lost. If it goes into either of the two cups flanking the centre one, then the 10 p is returned. If it enters an outer cup, then 30 p is returned. What are the chances of winning or losing at this game?

On leaving the first junction the ball has an equal probability of going to the left or right, so the probabilities of arriving at each of these junctions is 0.5. Now consider a ball that reached the left-hand junction. It can go left or right, and if it goes left, then it will reach the left-hand junction of the second row: the one containing three junctions. For this to happen, two independent events must take

71

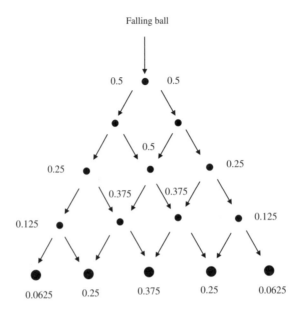

Fig. 7.1.   The pinball machine.

place: it must have gone left at two junctions with a probability 0.5 that it will do so each time. Hence the probability that it will get there is: $0.5 \times 0.5 = 0.25$, as marked in Fig. 7.1. By symmetry, this is also the probability of reaching the right-hand junction of that row. To get to the middle junction of the row requires either of the combinations left-right or right-left, each of which has a probability $0.5 \times 0.5 = 0.25$. Since the combinations left-right and right-left are mutually exclusive, the probability of reaching the middle junction is obtained by adding the two probabilities, i.e., $0.25 + 0.25 = 0.5$. These three probabilities are marked along the second row of junctions and, of course, they add to unity, since the ball must arrive somewhere on that row.

Similar reasoning gives the probabilities marked along the third row of junctions and along the cups. It will be seen that the game is superficially attractive to players. They lose their money on $0.375 \left(= \frac{3}{8}\right)$ of the occasions they play, but on the other 0.625 of the occasions they either get their money back or make a handsome profit. Let us see what their chances really are. The probability of losing is $0.375 \left(= \frac{3}{8}\right)$, of getting their money returned is: $0.25 + 0.25 = 0.5 = \frac{1}{2}$ and of getting three times their stake returned is $0.0625 + 0.0625 = 0.125 = \frac{1}{8}$. So the expectation is that if they play eight games their return will be zero on three games, 10 p on four games and 30 p on one game: a return of 70 p in all. However, they inserted 80 p to play the eight games, so they are a net loser. Of course, if their luck is in and they quit when they are ahead, then they can win something, but if they play the game a sufficiently long time, they are almost bound to lose.

This is an example where we have considered combinations of probabilities, some independent and some mutually exclusive. As an example, we can consider the ways the ball can get to the centre cup in Fig. 7.1. Representing a left deflection by L and a right deflection by R, there are six mutually exclusive routes the ball can take:

$$\text{LLRR} \quad \text{LRLR} \quad \text{LRRL} \quad \text{RRLL} \quad \text{RLRL} \quad \text{RLLR}$$

Each step, to left or right, is independent of the previous ones, so the probability of each path is:

$$0.5 \times 0.5 \times 0.5 \times 0.5 = 0.0625,$$

but since the routes are mutually exclusive, the total probability of reaching the central cup is:

$$6 \times 0.0625 = 0.375.$$

Probability machines that are much more complicated can be designed with many more rows of junctions, and with the junctions, which may be a simple pin off which the ball bounces, so designed that the chances of going left or right are not equal. However, no matter how complicated the probability machine, the chances of ending in each of the cups can be calculated using the principles explained above.

7.2.　Pontoon: A Card Game

This game, called Pontoon in the United Kingdom, is an extremely corrupted form of the French name *Vingt-et-un*, and is called Blackjack in the United States. It is a popular casino gambling game. Cards are associated with a number of points, with the cards 2 to 10 giving the number they display. Court cards, i.e., a jack, queen, or king, count for 10 points, while an ace has a degree of flexibility and may be counted by players as either 1 point or 11 points, according to which best serves their requirements.

The dealer initially deals two cards to each player and to herself. The players then request further cards, one at a time and up to a maximum of three further cards, by saying "twist" or "hit", with the objective of getting as close as possible to 21 without exceeding 21. When a player wishes to stop receiving more cards, he says "stick" or "stand". If his total exceeds 21 he goes "bust": he loses that hand and forfeits his stake money to the dealer. Thus there is a measure of judgement to be exercised, although luck also plays a major role. For example, if the player has three cards, $2 + 6 + 8$, giving a total of 16 points, and then requests a further card that turns out to be a 6 or greater, then he will lose that hand. When all the players have received the cards they requested, the dealer reveals her two cards and adds to them until she wishes to stop. If she goes bust, then she pays all players who did not go bust an amount equal to that they staked. If she does not go bust, then she pays out to all those with a higher number of points but takes the stakes of all those with less than *or equal* to her number of points.

The above is a basic description but there are extra aspects. If 21 points are obtained with two cards, which must be an ace, plus a 10 or a court card, then that is called "pontoon" and will beat any other hand. Another good combination is a "five-card trick", which is five cards totalling less than 21, which beats all other hands except a pontoon. What has been given is not a complete description of the game, but suffices for present purposes. The dealer's advantage in this game is first, that if she has a hand *equal* in value to that of another player, then she wins and, second, that she knows something about what the other players have, for example, the number of cards she dealt them, although not their total of points.

A complete statistical analysis of Pontoon is possible but complicated — just think of the number of different ways of achieving a five-card trick. There are some hands for which it is fairly simple to estimate the probabilities. For example, we consider the probability of getting 19 with two cards. One way is to get 10 points +9 points. The 10 may be obtained with a 10 or a court card and the probability of getting one of these from the pack is $\frac{16}{52}$. Now there are 51 cards left in the pack, of which 4 are 9s. Hence the probability of getting a 9 as a second card is: $\frac{4}{51}$. The selections are independent, so the chance of getting first 10 points and then 9 is:

$$P(10, 9) = \frac{16}{52} \times \frac{4}{51} = 0.02413.$$

The probabilities of this and other selections are now listed:

$$P(10, 9) = \frac{16}{52} \times \frac{4}{51} = 0.02413$$

$$P(9, 10) = \frac{4}{52} \times \frac{16}{51} = 0.02413$$

$$P(11, 8) = \frac{4}{52} \times \frac{4}{51} = 0.00603$$

$$P(8, 11) = \frac{4}{52} \times \frac{4}{51} = 0.00603$$

The last two ways of getting 19 involve having an ace and 8 either way round. These four combinations are mutually exclusive, so the total probability of getting 19 points with two cards is:

$$P(19) = 0.02413 + 0.02413 + 0.00603 + 0.00603 = 0.06032,$$

or about 1 chance in 16. Of course, one may get 19 with three or four cards, so the total probability of getting 19 in a hand of Pontoon is higher than this. If 19 is obtained with five cards, then that is a five-card trick and is a stronger hand than just getting 19 points.

In the casino game of Pontoon, or Blackjack, the cards are peeled one by one from a deck of cards placed face downwards. There are professional gamblers who "count the cards", that is, they mentally record aspects of the cards they have already seen so as to reassess the probabilities for future cards. As a simple example, if the first 20 cards dealt from the pack contained one ace, then the probability that the next card will be an ace is increased from $\frac{4}{52} = 0.07962$, the probability before any cards are dealt, to: $\frac{3}{32} = 0.09375$. Such information can, in the hands of a clever gambler, slant the odds in their favour to the extent of wiping out the dealer's advantage and substituting a player's advantage. Casino managers are always on the lookout for card counters, and they are excluded from casinos when they are detected.

7.3.   The Thrower's Chance of Winning at American Craps

There are many different routes by which the thrower may win, or lose, in the American Craps dice game. The game can be won or lost in a single throw: getting a "natural" to win or by getting 2 (Fig. 7.2), 3 or 12 to lose. However, there are more complicated ways of winning and losing. We shall find the combined probability of all

Just my luck: snake eyes!

Fig. 7.2.   Two ways of losing playing American craps!

the winning ways; they are mutually exclusive so we need to identify all the routes, find their probabilities, and add them together.

Throwing a "natural"

The probability of this was found in Section (2.3) as: $P_{natural} = \frac{2}{9}$ $= 0.22222$.

Setting a "point" of 4 (or 10) and then winning

In Section (2.3) the probability of obtaining 4 was found as $\frac{1}{12}$. Having obtained a 4, the thrower now embarks on a series of throws. The only throws relevant to winning or losing after setting this point are getting 4 again to win, with a probability of $\frac{1}{12}$ on each throw, or getting 7 to lose with a probability of $\frac{1}{6}$ on each throw. The chance of getting a 4 before getting a 7 is:

$$\frac{\text{probability of 4}}{\text{probability of 4} + \text{probability of 7}} = \frac{\frac{1}{12}}{\frac{1}{12} + \frac{1}{6}} = \frac{1}{3}. \qquad (7.1)$$

Similarly, the probability of getting a 7 before getting a 4 is:

$$\frac{\text{probability of 7}}{\text{probability of 4} + \text{probability of 7}} = \frac{\frac{1}{6}}{\frac{1}{12} + \frac{1}{6}} = \frac{2}{3},$$

but we are not directly interested in this probability because we are finding the total probability of winning. Winning by this route has two independent components: first establishing a "point" of 4 and subsequently getting a winning throw of 4. The composite probability of this is:

$$P_4 = \frac{1}{12} \times \frac{1}{3} = \frac{1}{36}. \qquad (7.2)$$

Since the probabilities are the same for a total of 10 we have:

$$P_{10} = P_4. \qquad (7.3)$$

Setting a "point" of 5 (or 9) and then winning

In Section (2.3) the probability of obtaining 5 was found as $\frac{1}{9}$. The only throws relevant to winning or losing after setting this point are getting 5 again to win, with a probability of $\frac{1}{9}$ on each throw, or getting 7 to lose, with a probability of $\frac{1}{6}$ on each throw. For the relevant throws the chance of winning is:

$$\frac{\text{probability of 5}}{\text{probability of 5} + \text{probability of 7}} = \frac{\frac{1}{9}}{\frac{1}{9} + \frac{1}{6}} = \frac{2}{5}. \qquad (7.4)$$

Winning by this route has two independent components: first establishing a "point" of 5 and subsequently getting a winning throw of 5. The composite probability of this is:

$$P_5 = \frac{1}{9} \times \frac{2}{5} = \frac{2}{45}. \qquad (7.5)$$

Since the probabilities are the same for a total of 9 we have:

$$P_9 = P_5. \qquad (7.6)$$

Setting a "point" of 6 (or 8) and then winning

In Section 2.3 the probability of obtaining 6 was found as $\frac{5}{36}$. The only throws relevant to winning or losing after setting this point are getting 6 again to win, with a probability of $\frac{5}{36}$ on each throw, or getting 7 to lose, with a probability of $\frac{1}{6}$ on each throw. For the relevant throws the chance of winning is:

$$\frac{\text{probability of 6}}{\text{probability of 6} + \text{probability of 7}} = \frac{\frac{5}{36}}{\frac{5}{36} + \frac{1}{6}} = \frac{5}{11}. \qquad (7.7)$$

The composite probability of first establishing a "point" of 6 and subsequently getting a winning throw of 6 is:

$$P_6 = \frac{5}{36} \times \frac{5}{11} = \frac{25}{396}. \qquad (7.8)$$

Since the probabilities are the same for a total of 8 we have:

$$P_8 = P_6. \tag{7.9}$$

Summing the probabilities of these mutually exclusive routes for winning, the overall probability of winning from the thrower is:

$$P_{win} = P_{natural} + P_4 + P_5 + P_6 + P_8 + P_9 + P_{10}$$
$$= \frac{2}{9} + \frac{1}{36} + \frac{2}{45} + \frac{25}{396} + \frac{25}{396} + \frac{2}{45} + \frac{1}{36} = 0.492929.$$

Thus the bank has a slight advantage which, in any profitable casino, is a necessary condition for running any game. As the game is played, the bank and the thrower stake an equal amount and the winner takes all. For every £500 staked by the bank another £500 will be staked by the thrower who will, on average, receive £493. Thus, on average, the bank receives £7 profit for every £500 invested: a return of 1.4% per game, which is small but almost assured over a long period.

Problems 7

7.1 A probability pinball machine, similar in principle to that described in Section (7.1), is illustrated below.

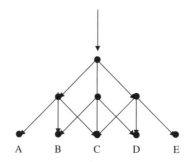

The ball has a probability of 1/3 of going along each of the channels shown. What is the probability of landing in each of the cups marked by a letter symbol? Analyse the profit or loss of using the machine if it costs 10 p per ball, no return was made on cup C, the 10 p was returned on cups B and D, and 20 p was returned on cups A and E.

7.2 What is the probability of obtaining a points total of 20 with two cards in a game of Pontoon?

7.3 In a game of American Craps, what is the probability that the thrower loses by:

(i) setting a point of 4 and then subsequently losing by throwing a 7?

(ii) setting a point and then subsequently losing by throwing a 7?

Chapter 8

The UK National Lottery, Loaded Dice and Crooked Wheels

Thou shalt not steal, an empty feat, When it's so lucrative to cheat
(Arthur Henry Clough, 1862, "The Latest Decalogue")

8.1.   The Need to Test for Fairness

In Section 3.1 we defined logical probability, which is based on reasoning and is heavily influenced by ideas of symmetry. We know that something is wrong when a coin comes down heads 100 times in 100 spins. But at what level should we be concerned: 90 times out of 100, 70 times out of 100, 60 times out of 100? After all, the coin may not have two heads but may be so unbalanced, or biased, that the mechanics of spinning it leads to a preponderance of heads. To see how to answer this question we first consider a completely different problem concerning outcome and expectation.

During the Second World War scientists in the United States were working on the development of a nuclear bomb. One problem they faced was to determine how neutrons would penetrate a barrier — a problem that they needed to solve quickly, but without any previous experience to guide them. A neutron interacting with an atomic nucleus in the barrier could be unaffected, absorbed or scattered. The probabilities of these possible interactions, which depended on the neutron energy, were known but this knowledge could not be converted into an analytical solution of the problem. The problem was eventually solved by two of the scientists working on the project, Stanislaw Ulam (1909–1984) and John von Neumann (1903–1957), by a novel numerical procedure. They could estimate how far a neutron would travel on average before encountering a nucleus, so they simulated the path of a neutron through a barrier. At each interaction they decided what would happen by the generation of a random number that would determine whether the atom was unaffected, absorbed or scattered and, if scattered, in which direction, and also how far it would travel to the next interaction. By following a large number of such simulated paths, they were able to determine the probability that a neutron would penetrate the barrier. The generation of a random number is similar to what happens in various gambling activities — throwing a die or spinning a roulette wheel, for example, so a procedure of this kind is called a *Monte Carlo method*. With the advent of computers that enable large numbers of Monte Carlo trials to be made, the method has been used in many areas, both in science and in the social sciences. However, it is important for the success of the method that the random numbers generated should be truly random.

Premium Bonds are a United Kingdom national institution that relies on the generation of random numbers. These bonds yield an average interest at a predetermined rate, but the interest is not distributed as income to each individual bond holder; it is used to provide prizes for a lottery based on the numbers of the bonds. The numbers of the bonds are generated each month by ERNIE, an acronym for Electronic Random Number Indicator Equipment. ERNIE uses the random noise produced in transistors to control an oscillator, the output of which is interpreted as numerical digits and letters. In this way, something of the order of one million random bond numbers can be generated per hour. This produces truly random numbers, since it is based on completely unpredictable physical events, but for some scientific purposes it would be unsuitable. For many scientific calculations truly random numbers are required but it is desirable that they should be generated in a reproducible way, so that the calculation that produces them can be rerun with the same numbers if necessary. For this reason, much effort has been expended in creating *pseudo-random number generators* that produce a series of random numbers by some repeatable mathematical process. Producing such generators is a surprisingly difficult thing to do and to be confident in the performance of a generator one must be able to test its output in some way.

8.2.   Testing Random Numbers

To illustrate the methods of testing for the fairness of a random number generator, we shall consider a generator that is intended to produce a string of random digits, from 0 to 9. The generator we test will be of the following form:

$$x_{n+1} = Frac\{(x_n + \pi)^2\}, \qquad (8.1a)$$

$$y_{n+1} = Int(10 \times x_{n+1}). \qquad (8.1b)$$

What this means will now be described. A series of fractional numbers is produced of the form:

$$x_1, x_2, x_3 ....$$

of which x_n is the nth one. Let this fractional number be 0.65382961. This number is added to the special number $\pi = 3.14159265$ to give 3.79542226. This number is then squared to give 14.40523013. That is what happens within the outer bracket in Equation (8.1a). The next thing that we do, represented by *Frac* in the equation, is to take the fractional part of that number, 0.40523013, which gives x_{n+1}. Now we move down to Equation (8.1b), where we multiply x_{n+1} by 10 to give 4.0523013 and then take the whole number part, represented by *Int* in the equation, to give the digit 4, which is y_{n+1}. To continue the series from x_{n+1} one can generate x_{n+2} and y_{n+2} and so on. To get started, it is necessary to provide an initial

fractional number, x_0, called the seed, for the random digit generator. A computer can readily generate a long string of digits in this
way. As long as the same computer is used, then the same string
of digits will be generated by the same seed. If a different set of
random digits is required, then it is only necessary to change the
seed.

A simple computer programme was used to generate 50 digits in this way and the numbers of each digit are shown in
Table 8.1.

The average times of occurrence of a digit is 5, but there is considerable variation about that average. Can we deduce just from
looking at the table that it is an unsatisfactory random digit generator? Not necessarily so. Inevitably, just by chance, the numbers of
each digit found will vary and, indeed, it would have been surprising — even suspicious — if we had found a row of ten 5 s. What
we need is an objective numerical measure to see if what we have
obtained could have come about with reasonable probability with
a good random number generator. To do this, we are going back to
the χ^2 test that was described in Section 6.2 to test the efficacy of a
drug. Table 8.1 represents the observed table with observations, O,
and Table 8.2 the expected table E.

Table 8.1. Observed table for 50 digits produced by the random digit
generator described in Equations (8.1).

Digit	0	1	2	3	4	5	6	7	8	9	Total
Number	2	8	10	4	2	1	7	7	6	3	50

Table 8.2. The expected table E for 50 digits with a uniform distribution.

Digit	0	1	2	3	4	5	6	7	8	9	Total
Number	5	5	5	5	5	5	5	5	5	5	50

Now we calculate χ^2, as we did in Equation (6.1), except that we now have 10 boxes rather than 4:

$$\chi^2 = \sum_{i=1}^{10} \frac{(O_i - E_i)^2}{E_i}$$

$$= \frac{(2-5)^2}{5} + \frac{(8-5)^2}{5} + \frac{(10-5)^2}{5} + \frac{(4-5)^2}{5} + \frac{(2-5)^2}{5}$$

$$+ \frac{(1-5)^2}{5} + \frac{(7-5)^2}{5} + \frac{(7-5)^2}{5} + \frac{(6-5)^2}{5} + \frac{(3-5)^2}{5}$$

$$= 16.4.$$

However, we cannot use Table 6.1 for our probabilities because we have a different number of *degrees of freedom* in the present case. You will recall that, in Section 6.2, the number of degrees of freedom was found as the number of arbitrary entries you could make in the table, given the sums in rows and columns. In the present case, with only one row, it is just the sum of the row, 50, that is relevant. There are 10 digit boxes and until you have filled in 9 of them the contents of the table are not defined, so in this case we have 9 degrees of freedom. Published tables give the value of χ^2 for various numbers of degrees of freedom and we show an excerpt for 9 degrees of freedom in Table 8.3.

Comparing the value of χ^2 with the entries of the table shows that there is only about a 1 in 20 chance that the value found for χ^2, or some greater value, could come about just by chance. This indicates that the random digit generator might be suspect.

Using the same random digit generator, the distribution for 1,000 digits is given in Table 8.4.

Table 8.3. Probabilities associated with the χ^2 distribution for 9 degrees of freedom.

Probability	0.50	0.30	0.20	0.10	0.05	0.025	0.02	0.01	0.005
χ^2		8.343	10.656	12.242	14.684	16.919	19.023	19.679	21.666	23.589

Table 8.4. Analysis of 1,000 digits produced by the random digit generator described in Equations (8.1).

Digit	0	1	2	3	4	5	6	7	8	9	Total
Number	109	99	123	91	99	83	89	103	98	106	1,000

Table 8.5. Analysis of 1,000 digits produced by a more sophisticated random fraction generator followed by Equation (8.1b).

Digit	0	1	2	3	4	5	6	7	8	9	Total
Number	99	94	98	104	92	85	102	101	115	110	1,000

The expected table would have 100 in each box, so for this distribution the value of χ^2 is:

$$\frac{(9)^2}{100} + \frac{(-1)^2}{100} + \frac{(23)^2}{100} + \frac{(-9)^2}{100} + \frac{(-1)^2}{100} + \frac{(-17)^2}{100} + \frac{(-11)^2}{100}$$

$$+ \frac{(3)^2}{100} + \frac{(-2)^2}{100} + \frac{(6)^2}{100} = 11.52.$$

This gives a better indication that the random digit generator may be reliable, since there is now about a 1 in 4 chance of getting this value of χ^2 or greater.

The step described in Equation (8.1a) generated a series of fractional numbers with a uniform distribution between 0 and 1, and there are much more complicated and effective processes for doing this. The result of producing 1,000 random digits, by using one of these generators for a uniform distribution of numbers in the range 0 to 1, much used by Monte Carlo modellers, followed by Equation (8.1b), gave the outcome shown in Table 8.5.

This distribution gives $\chi^2 = 6.76$ and from Table 8.2 we see that there is a somewhat greater than 50% chance that random fluctuations would give this value of χ^2 or greater. One could have more confidence in this generator than the one first described.

The general principles described here for testing random digit generators can also be used for testing the output of the various pieces of equipment used for gambling.

8.3.   The UK National Lottery

The way that the United Kingdom National Lottery operates was described in Section 4.4. Balls numbered from 1 to 49 are selected at random from a rotating drum that ejects them individually without human intervention. In this way, it is expected that the numbers are produced without any bias. To see if this is so, we look at how frequently the numbers 1 to 49 have been picked and Table 8.6 gives the relevant frequencies for the whole history of the lottery up to 4th October 2006.

The first thing we notice about this table is the large range of frequencies, from 108 for number 20 to 167 for number 38. This might lead us to wonder whether this is an indication that the selection of numbers was flawed in some way. To check this, we need to do a χ^2 test. The average number of times each number was selected is 138.4; while this average is fractional, and hence not a possible frequency for selecting a number, for the purpose of working out the

Table 8.6.   The frequency of UK lottery winning numbers up to 4th October 2006 (UK National Lottery).

(1) 134	(2) 138	(3) 133	(4) 130	(5) 131	(6) 145	(7) 147
(8) 133	(9) 144	(10) 142	(11) 151	(12) 141	(13) 115	(14) 132
(15) 125	(16) 123	(17) 132	(18) 136	(19) 146	(20) 108	(21) 125
(22) 137	(23) 152	(24) 126	(25) 155	(26) 138	(27) 143	(28) 141
(29) 137	(30) 146	(31) 149	(32) 145	(33) 147	(34) 130	(35) 141
(36) 127	(37) 127	(38) 167	(39) 129	(40) 148	(41) 111	(42) 137
(43) 158	(44) 159	(45) 141	(46) 134	(47) 156	(48) 153	(49) 135

value of χ^2 it will be taken as the expected value, E, to be inserted in Equation (6.1). There are 49 terms in the calculation of χ^2 and we indicate the first two and last two of these in (8.2):

$$\chi^2 = \frac{(134-138.4)^2}{138.4} + \frac{(138-138.4)^2}{138.4} \cdots + \frac{(153-138.4)^2}{138.4}$$

$$+ \frac{(135-138.4)^2}{138.4} = 51.56. \tag{8.2}$$

The total number of numbers picked in the lottery up to 4th October 2006 was 6,780 and, given this total, 48 of the frequencies must be given to define the table, meaning that there are 48 degrees of freedom in this significance test. Commonly available χ^2 tables do not normally give probabilities for 48 degrees of freedom, but there are Web sources that enable such probabilities to be derived. A partial table for 48 degrees of freedom is given in Table 8.7.

From the table, we can see that the probability of getting $\chi^2 = 51.56$, or some greater value, is greater than 0.25, so we have no reason to suppose that the selection of National Lottery numbers is biased in any way. This example shows the importance of doing objective statistical tests before jumping to conclusions on the basis of just examining the numbers involved. Without doing such a test it might be concluded that the discrepancy between the number of times that 38 and 20 occurred indicated some failure of the number selection system. However, the χ^2 test confirms that this is not so, and such variation is within the bounds of normal random fluctuation.

Table 8.7. Probabilities associated with the χ^2 distribution for 48 degrees of freedom.

Probability	0.50	0.25	0.10	0.05	0.025	0.01
χ^2	47.33	54.20	60.91	65.17	69.02	73.68

8.4.   American Craps with Loaded Dice

In Section 7.2 it was shown that the thrower's chance of winning for the American Craps dice game is 0.492929, just slightly under 0.5, so that the bank makes a small, but assured, profit in the longer-term. However, let us suppose that the dice are slightly loaded. This could be done by inserting a small lead weight within each die, as shown in Fig. 8.1, on the line between the centre of the 1-face and the 6-face (which are opposite each other).

Now, when the die rolls, it will have a slight tendency to end up with the lead weight as low as possible, i.e., closest to the table, thus increasing the chance of getting a 1 and decreasing the chance of getting a 6. Now we suppose that instead of the six numbers each having a probability of $\frac{1}{6} = (0.1666^*)$ the probabilities are:

$$P(1) = 0.175000, \ P(2) = P(3) = P(4) = P(5) = \frac{1}{6}$$

and $P(6) = 0.158333.$

How much difference would this make?

To determine the new probabilities of winning and losing we must go through the analysis, as was done in Section 7.2. Since the probability is not the same for all faces the analysis is a little more complicated, but the principles involved are the same. In each case the probability of getting a particular combination of two numbers

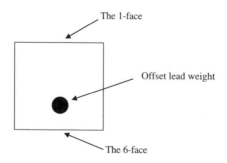

Fig. 8.1.   A loaded die.

on the two dice is just the product of the individual probabilities for each of the numbers.

Probability of a "natural"

This requires one or other of the combinations shown with their probabilities as indicated:

Getting 7:

$$6+1 \quad 1+6 \quad 2 \times (0.1750 \times 0.158333) = 0.055417$$

$$5+2 \quad 4+3 \quad 3+4 \quad 2+5 \quad 4 \times \left(\frac{1}{6} \times \frac{1}{6}\right) = 0.111111$$

Total probability of 7: $P_7 = 0.055417 + 0.111111 = 0.166528$.

We notice that this is less than $\frac{1}{6}$ ($= 0.166667$), the probability with unleaded dice. In all the following calculations of probability with loaded dice we give the corresponding result with fair dice in parentheses.

Getting 11:

$$6+5 \quad 5+6 \quad 2 \times \left(0.158333 \times \frac{1}{6}\right) = 0.052778 \ (0.055556)$$

The probability of a "natural" is $P_{natural} = 0.166528 + 0.052778 = 0.219306 \ (0.222222)$.

Setting a "point"

Setting a "point" of 4 and then winning

$$3+1 \quad 1+3 \qquad 2 \times \left(\frac{1}{6} \times 0.1750\right) = 0.058333$$

$$2+2 \qquad \frac{1}{6} \times \frac{1}{6} = 0.027778$$

Hence: $P_4 = 0.058333 + 0.027778 = 0.086111 \ (0.083333)$.

The probability of then getting 4 before 7 is:

$$\frac{P_4}{P_4 + P_7} = \frac{0.086111}{0.086111 + 0.166528} = 0.340846 \ (0.333333).$$

To win with a "point" of 4 requires the two steps: first, to establish the "point" and then to win with it. The probability of taking these two independent steps is:

$$P_{w4} = 0.086111 \times 0.340846 = 0.029351 \ (0.027778).$$

Setting a "point" of 5 and then winning

$$4 + 1 \quad 1 + 4 \qquad 2 \times \left(\frac{1}{6} \times 0.1750 \right) = 0.058333$$

$$2 + 3 \quad 3 + 2 \qquad 2 \times \left(\frac{1}{6} \times \frac{1}{6} \right) = 0.055556$$

Hence: $P_5 = 0.058333 + 0.055556 = 0.113889 \ (0.111111).$
    The probability of then getting 5 before 7 is:

$$\frac{P_5}{P_5 + P_7} = \frac{0.113889}{0.113889 + 0.166528} = 0.406142 \ (0.400000).$$

The probability of taking the two independent steps is:

$$P_{w5} = 0.113889 \times 0.406142 = 0.046255 \ (0.044444).$$

Setting a "point" of 6 and then winning

$$5 + 1 \quad 1 + 5 \qquad 2 \times \left(\frac{1}{6} \times 0.1750 \right) = 0.058333$$

$$4 + 2 \quad 2 + 4 \quad 3 + 3 \qquad 3 \times \left(\frac{1}{6} \times \frac{1}{6} \right) = 0.083333$$

Hence: $P_6 = 0.058333 + 0.083333 = 0.141666 \ (0.138889).$

The probability of then getting 6 before 7 is:

$$\frac{P_6}{P_6 + P_7} = \frac{0.141666}{0.141666 + 0.166528} = 0.459665 \ (0.454545).$$

The probability of taking the two independent steps is:

$$P_{w6} = 0.141666 \times 0.459665 = 0.065119 \ (0.063131).$$

Setting a "point" of 8 and then winning

$$6+2 \quad 2+6 \qquad\qquad 2 \times \left(\frac{1}{2} \times 0.158333 \right) = 0.052778$$

$$5+3 \quad 3+5 \quad 4+4 \qquad 3 \times \left(\frac{1}{6} \times \frac{1}{6} \right) = 0.083333$$

Hence: $P_8 = 0.052778 + 0.083333 = 0.136111 \ (0.138889).$
The probability of then getting 8 before 7 is:

$$\frac{P_8}{P_8 + P_7} = \frac{0.136111}{0.136111 + 0.166528} = 0.449747 \ (0.454545).$$

The probability of taking the two independent steps is:

$$P_{w8} = 0.136111 \times 0.449747 = 0.061216 \ (0.063131).$$

Setting a "point" of 9 and then winning

$$6+3 \quad 3+6 \qquad\qquad 2 \times \left(\frac{1}{6} \times 0.158333 \right) = 0.052778$$

$$5+4 \quad 4+5 \qquad\qquad 2 \times \left(\frac{1}{6} \times \frac{1}{6} \right) = 0.055556$$

Hence: $P_9 = 0.052778 + 0.055556 = 0.108334 \ (0.111111).$

The probability of then getting 9 before 7 is:

$$\frac{P_9}{P_9 + P_7} = \frac{0.108334}{0.108334 + 0.166528} = 0.394140 \ (0.400000).$$

The probability of taking the two independent steps is:

$$P_{w9} = 0.108334 \times 0.394140 = 0.042699 \ (0.044444).$$

Setting a "point" of 10 and then winning

$$6 + 4 \quad 4 + 6 \qquad 2 \times \left(\frac{1}{6} \times 0.158333 \right) = 0.052778$$

$$5 + 5 \qquad \frac{1}{6} \times \frac{1}{6} = 0.027778$$

Hence: $P_{10} = 0.052778 + 0.027778 = 0.080556 \ (0.083333).$
The probability of then getting 10 before 7 is:

$$\frac{P_{10}}{P_{10} + P_7} = \frac{0.080556}{0.080556 + 0.166528} = 0.326027 \ (0.333333).$$

The probability of taking the two independent steps is:

$$P_{w10} = 0.080556 \times 0.326027 = 0.026263 \ (0.027778).$$

The overall probability of the thrower winning with the loaded dice is thus:

$$P_w = P_{natural} + P_{w4} + P_{w5} + P_{w6} + P_{w8} + P_{w9} + P_{w10} = 0.490209.$$

While this may not seem very different from the probability of winning with the fair dice, 0.492929, it is, in fact, a very important difference. Now, for every £500 staked by the thrower, he gets a return of £490, so that the banker makes a profit of £10 on a stake of £500: a return of 2.0% per game rather than the 1.4% received with the fair dice. A 43% increase in the rate of return is by no means trivial!

8.5. Testing for a Loaded Die

The effect of slightly loading the dice in a crap game, so that a 1 appearing is very slightly favoured, may not be very noticeable in the duration of a game involving 100 or 200 throws. However, as we have seen, it can give a substantial proportional increase in the profit to the bank. If it were suspected that the dice were loaded, then how could it be tested? The obvious way would be to throw the dice a large number of times to check that all the numbers appeared with equal frequency. The problem with this is that, even with a fair die, there would be some random fluctuations in the frequencies, as was seen in Table 8.4, that generated random digits with a perfectly fair random digit generator. To be sure that the dice was loaded we would have to carry out an objective test, like the χ^2 test. We would start with the *null hypothesis* that the die was a fair one, throw it many times, find the frequencies of the numbers 1 to 6, and then check that the distribution found is not an improbable one for a fair die.

To simulate what would happen if the loaded die was thrown many times, we are going to use a high quality random number generator that gives a uniform distribution between 0 and 1. Each number obtained is interpreted as a throw of the die in the following way:

between 0.000000 and 0.175000 is equivalent to throwing a 1 (probability 0.175000),

between 0.175000 and 0.341667 is equivalent to throwing a 2 (probability 0.166667),

between 0.341667 and 0.508333 is equivalent to throwing a 3 (probability 0.166666),

between 0.508333 and 0.675000 is equivalent to throwing a 4 (probability 0.166667),

between 0.675000 and 0.841667 is equivalent to throwing a 5 (probability 0.166667),

between 0.841667 and 1.000000 is equivalent to throwing a 6 (probability 0.158333).

This gives a probability of 0.175000 of obtaining a 1, a probability of 0.1583333 for a 6, and a probability of $\frac{1}{6}$ (very nearly) for the other numbers — the loading of the die considered in Section 8.3.

When 600 random numbers were generated, and translated into equivalent throws of the die, the result is as given in the first numerical row of Table 8.8.

For the null hypothesis of a fair die, the expected frequency of each number, shown in the column headed E, is 100. The actual numbers vary between 81 and 114. We can calculate the value of χ^2 as we did for the random digits in Section 6.2 but now, with 6 faces of the die rather than 10 digits, the number of degrees of freedom is 5. Part of a probability table for 5 degrees of freedom is given in Table 8.9.

The value of χ^2 for 600 throws is:

$$\chi^2 = \frac{(114-100)^2}{100} + \frac{(93-100)^2}{100} + \frac{(100-100)^2}{100} + \frac{(81-100)^2}{100}$$

$$+ \frac{(107-100)^2}{100} + \frac{(105-100)^2}{100} = 6.80.$$

Table 8.8.   A simulation of throwing a loaded die many times.

Total throws	E	1	2	3	4	5	6	χ^2
600	100	114	93	100	81	107	105	6.80
3,000	500	551	493	487	498	504	467	7.86
15,000	2,500	2,628	2,471	2,505	2,510	2,512	2,374	13.35
45,000	7,500	7,832	7,513	7,536	7,485	7,508	7,126	33.58

Table 8.9.   Probabilities associated with the χ^2 distribution for 5 degrees of freedom.

Probability	0.50	0.30	0.20	0.10	0.05	0.025	0.02	0.01	0.005
χ^2		4.351	6.064	7.289	9.236	11.070	12.832	13.388	15.086	16.750

From Table 8.9 we see that there is about a 25% chance of getting this value of χ^2, or some greater value, with a fair die; there is no evidence that the die is biased for a test with this number of throws. Now suppose that we increase the number of test throws to 3,000. The number 1 now appears clearly more frequently than any other but the value of χ^2, 7.86, or some greater value, has a probability of somewhat less than 0.2 of being found with a fair die. We could still not be confident that the die was unfair with this test.

When we do a test with 15,000 throws the evidence for a loaded die becomes stronger. The value of χ^2, 13.35, or some greater value, has a 1 in 50 chance of happening with a fair die. We may be fairly confident that the die is loaded. However, with 45,000 throws and a value of $\chi^2 = 33.58$ we may be completely certain that the die is loaded.

With 45,000 throws the proportion of throws for the various numbers are:

1   0.1740   2   0.1670   3   0.1675   4   0.1663   5   0.1668

6   0.1584

which are very close to expectation. For a large-scale test it is possible to find not only that the die is loaded, but also a good estimate of the way that it is loaded.

8.6.   The Roulette Wheel

A roulette wheel, as found in European casinos, contains 37 numbered slots uniformly distributed around its rim within an annular valley. When the wheel is spinning quite quickly, a steel ball is tossed into the valley after which it bounces around in haphazard fashion until eventually it lands in one of the slots. The slots are numbered from 0 to 36 and there are various ways of betting on the spin of the wheel. One may pick one of the 36 non-zero numbers and, if the

Fig. 8.2.   An American roulette wheel with 0 and 00.

ball lands on that number, the punter gets odds of 35:1, i.e., he gets 36 times his stake returned. Another way is to bet on either even or odd numbers, where the odds are 1 to 1, or evens, or on a group of six numbers where the odds are 5 to 1. If the only numbers on the wheel were 1 to 36, then the casino and the punter would be playing on even terms. However, on average, once every 37 spins of the wheel, the ball lands in the 0, and then the casino takes all the stakes; this is its profit.

The standard roulette wheel used in the United States is somewhat more profitable to the casino, because it not only has a 0 slot, but also a 00 slot when the casino takes all (Fig. 8.2).

A wheel which is slightly off-centre might give an increased tendency for the ball to fall in the 0. The ball is usually made of steel, so a bias towards 0 may also be made with a magnet, but this form of cheating is so easily detected that it is never used. To test a roulette wheel for fairness, it is necessary to spin it many times. From the point of view of fairness, there are only two types of number: for a European wheel 0 and non-0. Suppose that a wheel gave a frequency of 0 s 10% greater than it should. Could this be detected with, say,

3,700 spins of the wheel? With the null assumption that the wheel is fair, the observed and expected tables are as shown below:

Observed

0	non-0
110	3,590

Expected

0	non-0
100	3,600

The value of χ^2 for these tables is 1.028 which, for one degree of freedom (Table 6.1), would not indicate bias with any certainty. However, if 0 turned up 1,100 times in 37,000 spins, the value of χ^2 would be 10.28; the chances of getting this value of χ^2, or a greater value, is only 1 in 1,000, so there would then be clear evidence for a biased wheel.

Where casinos are licensed to operate, either locally or nationally, they are also monitored and the punter may be reasonably sure that the equipment used is fair and without bias.

Problems 8

8.1 When a die is thrown 600 times the occurrence of the various sides is as follows:

1	2	3	4	5	6
112	113	81	109	101	84

On the basis of this test is there a greater than 90% chance that the die is unfair?

(*Note*: The wording of this question implies that the probability of getting the associated value of χ^2, or some greater value, with a fair die is less than 10%.)

8.2 A heavily-biased die has the following probabilities of obtaining the various sides:

$$
\begin{array}{cccccc}
1 & 2 & 3 & 4 & 5 & 6 \\
\dfrac{1}{4} & \dfrac{1}{6} & \dfrac{1}{6} & \dfrac{1}{6} & \dfrac{1}{6} & \dfrac{1}{12}
\end{array}
$$

What is the probability that the thrower will get a natural?

8.3 In 370 spins of a European roulette wheel, a 0 occurs 20 times. Would you be justified in suspecting that the wheel was biased?

Chapter 9

Block Diagrams

Variety's the very spice of life, That gives it all its flavour
(William Cowper, 1785, "The Task, Book II: The Timepiece")

9.1. Variation in Almost Everything

Most natural classes of objects show variations. There is the saying: "They are as alike as two peas in a pod" but when you examine the peas in a pod carefully they are all different: in size, weight or blemishes on their surfaces. In particular, it is a matter of everyday experience that all human beings are different. Identical twins come from the same divided ovum and are thus identical in their DNA composition but, nevertheless, even the most similar identical twins differ in some small characteristics. The fact is that we are all the products both of nature and nurture and, whatever our genetic inheritance, the environment in which we develop and live, and our life experiences, will influence our characteristics. An individual may be genetically inclined to be tall, but if they have an inadequate diet, then they will never achieve the capability given by their genes. Differences between people can be both physical and non-physical so that, for example, people have wide differences of temperament. One person is placid, another excitable, one may be of happy disposition and another morose. As William Cowper stated in the quotation that begins this chapter, this variety is all part of the rich tapestry of life.

Non-physical characteristics can usually be expressed in a qualitative way but may be difficult or impossible to quantify.

Joe may be impulsive, Jim may be reflective and deliberate in his actions, and John somewhere in-between, but how can these characteristics be numerically described? By contrast, physical characteristics are easily measured. We may generally describe people as *tall*, *short* or of *average height*, but we can be more specific by giving their actual measured heights. Thus a man who is 1.91 metres (6′ $3\frac{1}{4}''$) tall would be regarded as tall in a European context, although he would be considered of normal height by the Watutsi people of Rwanda, many of whom grow to a height of 2.1 metres (nearly 7′) or more. This example gives us a warning. Comparative terms such as *tall* or *short* only have meaning when the context in which they are applied is known. A short man in Europe may be average, or even tall, by the standards of the pigmy people of the Congo, and *tall* in Europe may not mean the same as *tall* in Rwanda. We should stick to numbers: these mean the same thing everywhere.

Variation occurs in entities other than human beings. In other animals, in plants and in geographical features, such as the length of rivers, in all these there is variation. Only scientists deal with objects that, by their very nature, are identical. All atomic particles of the same kind are precisely the same. There is no way that two electrons can be distinguished — if there were, then the whole of physics as a subject would be radically different and the physical world would be very difficult, or even impossible to describe in a systematic way. But atomic particles are things we cannot see, although we can infer their existence with complete certainty. In the things that we *can* see, no two things are alike; however similar they may seem superficially, at some level of refinement they will be found to be different. Usually, variations are just matters of note but, particularly in the human context, they can be of greater importance, and may even be of commercial significance.

9.2.   A Shoe Manufacturer

The objective of a shoe manufacturer is to make shoes to be sold at a profit to a shoe wholesaler who will, in turn, sell them to a retailer at a profit who will, in turn, sell them to a customer at a profit. The last link in this chain is the customer, and satisfying the needs of customers is the key to the success of the whole enterprise. If a customer comes into a shop and cannot find the size of shoe to fit him or her, then a sale is lost. Another adverse factor in the profit-making enterprise would be if a large surplus of some particular sizes of shoe were to build up at any stage of the chain, because this would correspond to an unproductive tie-up of capital. It is of interest to all that the number of shoes made in a particular size should approximately match the number of customers needing that size.

Let us consider men's shoes. If it were possible to check the shoe size of every man in the community, then the proportions of shoes required of each size would be known. Although that would not be a feasible exercise, there is a process called *sampling*, that is described more fully in Chapter 14, by which a close approximation to the relative numbers required of each size can be found. Such a sampling exercise would probably find that the great majority of men's shoes fall within the United Kingdom sizes 6–13, with a tiny fraction outside that range. Extreme sizes, at least in the upward direction, would fall in the province of shops dealing with unusual clothing requirements — and the cost to customers of providing for such a need is usually quite high. However, within the normal commercial range, the obviously efficient thing to do would be to manufacture numbers of men's shoes with sizes in the same proportions as occur in the adult male population. Notional proportions, somewhat idealized, are shown in Table 9.1 for shoe sizes 6–13 in steps of $\frac{1}{2}$.

Table 9.1 gives the essential information about the proportions of shoes of each size required, but the detailed numbers must be studied at some length to appreciate the variation of demand over the size range. A visual, and better, picture showing the overall pattern of demand is given in Fig. 9.1, where the same information is presented in the form of a block diagram, or *histogram*, in which the height of a block gives the proportion of shoes required of the indicated size.

Table 9.1. Notional proportions of men's shoe sizes between 6 and 13.

Shoe size	Proportion	Shoe size	Proportion
6	0.0004	10	0.1761
$6\frac{1}{2}$	0.0022	$10\frac{1}{2}$	0.1210
7	0.0088	11	0.0648
$7\frac{1}{2}$	0.0270	$11\frac{1}{2}$	0.0270
8	0.0648	12	0.0088
$8\frac{1}{2}$	0.1210	$12\frac{1}{2}$	0.0022
9	0.1761	13	0.0004
$9\frac{1}{2}$	0.1995		

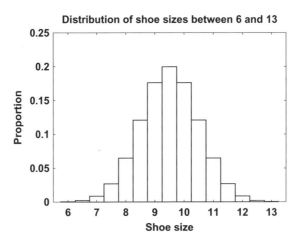

Fig. 9.1.   The distribution of men's shoe sizes shown as a histogram.

The symmetrical bell-shaped distribution shown in Fig. 9.1 is very common in naturally occurring quantities. Since values at either the extreme smaller end of the scale or at the extreme larger end are usually rare, a bell-shape is very common, although it may not necessarily be symmetrical. However, there are exceptions to the conditions that give bell-shaped distributions, and other types of distribution are possible.

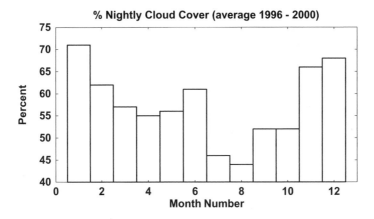

Fig. 9.2. Average cloud cover on a monthly basis at Trenton, Ontario (NOAA).

9.3.   Histogram Shapes

As an example of a non-bell-shaped histogram, Fig. 9.2 gives the average night-time cloud cover, month-by-month, averaged over a five year period at Trenton in Ontario — something of interest to astronomers.

Although this seems very different from the histogram representing shoe sizes, in fact, there is some similarity. The months of the year run cyclically and deciding to begin the year with the month of January is an arbitrary choice. If we re-plotted Fig. 9.2, starting with August as the first month, then the cloud cover steadily rises each month up to January, then falls steadily until April. Enhanced cloud cover in May and June spoils the bell-shaped pattern. In fact, we can interpret the cloud-cover histogram as a *bimodal distribution*, i.e., one with two peaks, occurring in January and June.

The bimodal nature of the cloud-cover distribution is not too pronounced, but there are situations where much stronger bimodal distributions can occur. One of the great wonders of nature, in Yellowstone Park in the United States, is the Old Faithful geyser

Fig. 9.3. The Old Faithful geyser, Yellowstone Park, USA (USGS).

(Fig. 9.3). This geyser may be faithful, but it is somewhat variable in its behaviour. It erupts at intervals between 65 and 92 minutes, mostly at the longer end, sending anything from 14,000 to 32,000 litres of boiling water to a height between 30 and 55 metres. The duration of each eruption is also very variable, being between 1.5 and 5 minutes. Figure 9.4 shows a histogram of the relative number of eruptions of various durations, with durations divided into intervals of $\frac{1}{3}$ minute. This means that the first interval is between 1.5 and 1.833 minutes, the second between 1.833 and 2.167 minutes and so on. In Figure 9.4 the bimodal nature of the frequency is clearly seen. Most eruptions last either about 2 minutes or 4 minutes, with 3-minute eruptions, or thereabouts, being comparatively rare.

Other shapes of histogram can occur, but only rarely. The bell-shaped and bimodal histograms account for most of the distributions that occur naturally.

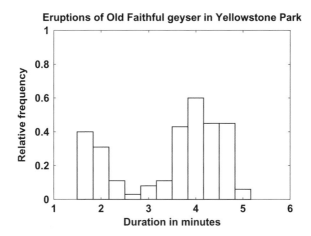

Fig. 9.4.   The distribution of eruption durations for Old Faithful.[a]

9.4.   Lofty and Shorty

In Fig. 9.1 a bell-shaped histogram comes from a quantity, shoe size, which is discrete, i.e., it can only occur with particular values. No shoes of size 9.35 are manufactured, and if you require shoes that are not one of the standard sizes, then you must go to the expense of having them made to measure. However, there are quantities that are continuously variable: height, weight etc., and we shall now see how to represent these by means of histograms. We have already introduced this idea for the eruption durations of Old Faithful, where the durations were divided into $\frac{1}{3}$ minute intervals.

It is common to give people nicknames dependent on their heights, especially in environments such as an army barracks, a school or a workplace. Thus, the 1.91 m (6' $3\frac{1}{4}$") man becomes known as "Lofty" and the 1.62 m man (5' 4") is called "Shorty". Sometimes, perversely and as a form of irony, the short man is called "Lofty" and the tall man "Shorty", but the main point here is that, in some way or other, the extremes of height are being highlighted in some

[a]B.W. Silverman, 1986, *Density Estimates for Statistics and Data Analysis*, Chapman and Hall.

way. The majority of men fall in the middle range of height and so they are not granted a distinguishing nickname relating to height — "Middley", for example, could be applied to most men.

If a man is asked for his height, he will usually quote it to the nearest centimetre or quarter-inch, but it is only an approximation. If someone says that he is 1.76 m tall, what he really means is that his height is closer to 1.76 m than to either 1.75 m or 1.77 m. In fact, in giving his height as he does, he is implying that his height is somewhere between 1.755 m and 1.765 m. Figure 9.5 shows the basis of this implication.

We now consider the question: "What is the probability that an adult British male, chosen at random, will have a height of 1.76 m?" The question, in a strictly scientific sense, is meaningless, since there may be nobody who is *exactly* 1.76 m tall, although there may be several men between 1.7599 m and 1.7601 m tall. The question only makes sense if we ask for the probability that the man will be in a certain range of height, for example, between 1.755 m and 1.765 m. If we could explore the whole population of British adult males and look at the proportions in various heights in ranges of 1 cm, we could produce a histogram looking something like Fig. 9.6.

We notice that the histogram shown in Fig. 9.6 is somewhat smoother in appearance than the one in Fig. 9.1, and this smoothness is dependent on the size of the interval we choose for the blocks. For example, in Fig. 9.7 we show the same distribution of heights but this time plotted in half-centimetre blocks, and the result is smoother than that shown in Fig. 9.6.

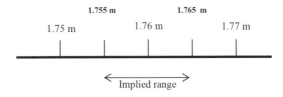

Fig. 9.5. The implied range of possible heights for a quoted height of 1.76 m.

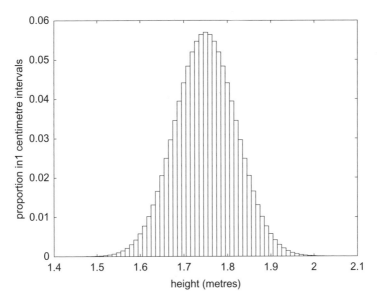

Fig. 9.6. A histogram showing notional proportion of UK males in 1 centimetre height intervals.

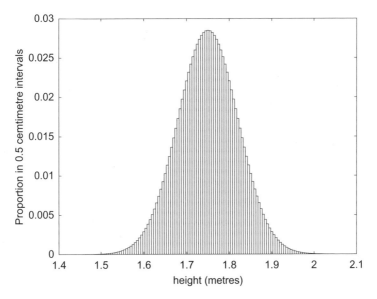

Fig. 9.7. A histogram showing notional proportion of UK males in 0.5 centimetre height intervals.

We notice that the heights of the blocks are less in Fig. 9.7 than in Fig. 9.6, since the blocks are narrower and the proportion of men with heights in a 0.5 centimetre interval centred on 1.75 m will obviously be one-half of the proportion in a 1 centimetre interval.

As we make the intervals smaller and smaller, so the saw-tooth serrations along the boundary get finer and finer. In the following chapter we shall see what happens when we take this to the limit of very tiny (theoretically zero) intervals.

Problem 9

9.1 Women's off-the-peg dresses in the UK are normally manufactured in sizes ranging from 6 to 26 in steps of 2. The proportions of women in each size category, excluding the very few outside that range, are:

6	0.07	18	0.09
8	0.10	20	0.04
10	0.15	22	0.03
12	0.20	24	0.02
14	0.16	26	0.01
16	0.13		

Draw a histogram to represent this table.

A shop has a policy of charging the same price for all sizes of a particular style. If it only stocked sizes 8 to 18, its overall costs would be reduced and its profit per garment sold would increase by 20%. In the interests of its shareholders should it reduce the range of sizes it sells?

Chapter 10

The Normal (or Gaussian) Distribution

... that of the strange and sinister embroidered on the very type of the normal and easy (Henry James, 1903, *The Ambassadors*)

10.1. Probability Distributions

Let us imagine that we take the distribution of men's heights, as illustrated in Figs. 9.6 and 9.7, and we make the intervals narrower and narrower, so that the appearance of the curve becomes smoother and smoother. Eventually we reach a distribution that is completely smooth, looking somewhat like the curve shown in Fig. 10.1. However, you may have noticed an important difference between this distribution and the ones shown in Figs. 9.6 and 9.7. When we went from wider to narrower blocks in going from Fig. 9.6 to Fig. 9.7, the lengths of the blocks went down, with the maximum length going from about 0.057 to about 0.028. You may well ask why it is that when we represent the situation of having infinitely narrow blocks, the maximum shoots up to nearly 7. How does this come about?

We can best understand this by going back to Fig. 9.1 and interpreting what we see there. The length of the block corresponding to size $9\frac{1}{2}$ is 0.1995, corresponding to the proportion of shoes required at that size. Another way of thinking about this is to say that, if we pick a pair of shoes at random from all the available pairs, then there is a probability of 0.1995 that it will be size $9\frac{1}{2}$. Similarly, the length of the block corresponding to size 10 (and also to size 9) is 0.1761, and this is the probability that a pair of shoes chosen at random will be size 10 (or size 9). We now ask the question: "What is the probability

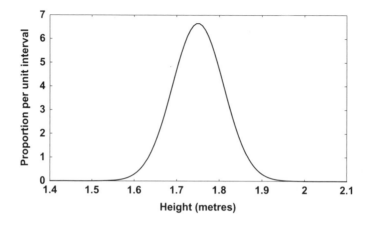

Fig. 10.1. A distribution of heights for infinitely small intervals.

that a pair of shoes chosen at random will be one of size 9, size $9\frac{1}{2}$ or size 10?" Since the choices are mutually exclusive — if one size is chosen, then the other two sizes are excluded — then the answer is obtained by adding the lengths of the blocks corresponding to the three sizes, i.e., $0.1761 + 0.1995 + 0.1761 = 0.5517$. Extending this idea, we now ask what the probability is of choosing a pair of shoes at random and having a size somewhere in the range 6–13. This is obtained by adding the lengths of all the blocks in Fig. 9.1 and the answer is found to be 1, which can be checked from the figures in Table 9.1. A probability of 1 means "certainty"; all the shoes are between sizes 6 and 13, so it is completely certain that the pair chosen will be one of these sizes.

With this interpretation of the lengths of the histogram blocks, we now consider Figs. 9.6 and 9.7, which portray the notional height distribution of UK males. The sum of the lengths of the blocks has to be 1 in each of the figures but, since there are twice as many blocks in Fig. 9.7 as there are in Fig. 9.6, the lengths of the blocks in Fig. 9.7 are, on average, one-half of the lengths of those in Fig. 9.6.

The continuous curve in Fig. 10.1, which contains no blocks of defined width with lengths to represent probabilities, has to be considered in a different way. In Fig. 10.2 we show a narrow strip of the curve, of thickness 0.01 m centred on the height 1.70 m.

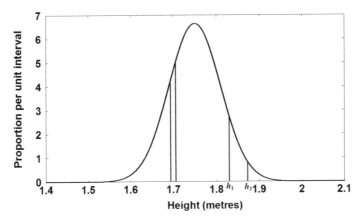

Fig. 10.2. A narrow strip of the distribution of Fig. 10.1, of width 0.01 centred on 1.70.

The strip covers the range of heights from 1.695 m to 1.705 m, and the probability that a man chosen at random would have a height between these limits is now taken as being given by the area of the strip, not the length of the strip as for the histograms. It follows from this that the probability that the randomly chosen man would have a height somewhere between h_1 and h_2, also shown in Fig. 10.2, is the total area under the curve between those two limits. Extending this idea, the probability that a man chosen at random would have a height between 1.40 m and 2.10 m is the total area under the curve and must be 1, or nearly so, because all, except a very few individuals, fall inside this range. To check this, Fig. 10.3 shows the curve covered by a mesh of blocks, each of area 0.1 (length 1 × breadth 0.1). Some blocks fall completely under the curve and some partially so, and within each block is given the estimated fraction of the block in the region below the curve to the nearest 0.1. Adding all these numbers gives 10.0 which, multiplied by 0.1, the area of each block, gives the area under the curve as 1, as it should be.

10.2.   The Normal Distribution

The symmetrical bell-shaped distribution shown in Fig. 10.1 is one that occurs frequently in nature. Its mathematical form

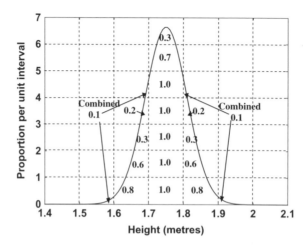

Fig. 10.3. The fractions of blocks under the curve judged to the nearest 0.1.

Fig. 10.4. Carl Friedrich Gauss, oil painting by Christian Albrecht Jensen.

was first described in detail and made well known by the German mathematician and scientist Johann Carl Friedrich Gauss (1777–1855), pictured in Fig. 10.4, who is usually known as Carl Gauss, since, for some reason, he discarded the "Johann" part of his given name. In fact, the distribution was first used in a mathematical context by Abraham de Moivre (1667–1754), a French mathematician, in 1733, but its widespread applications were not realized at that time.

This distribution is now called a normal distribution or, especially by physicists, a Gaussian distribution, and its properties are of great interest to the statistician. Because of its symmetrical form, it is clear that the peak of the curve represents the average, or mean, value. What is also of interest is the spread of the curve. To illustrate this, in Fig. 10.5 three normal curves with the same mean but with different spreads are shown.

Since the area under all the curves equals 1, this means that the greater the lateral spread, the less the height of the curve is — as is seen in the figure. We now have to find out how it is that different curves, as shown in Fig. 10.5, can all be described as normal curves.

10.3. The Variance and Standard Deviation

Clearly, what is needed to describe the difference between the curves in Fig. 10.5 is some numerical quantity related to their spread of values. To see what this could be, we consider the following two

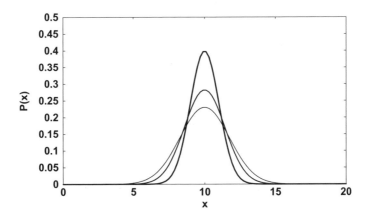

Fig. 10.5.   Three normal distributions with the same mean but different spreads.

sets of seven numbers:

$$
\begin{array}{llllllll}
\text{Set A} & 7 & 8 & 9 & 10 & 11 & 12 & 13 \\
\text{Set B} & 1 & 6 & 9 & 10 & 11 & 14 & 19
\end{array}
$$

The sum of each set of numbers is the same, 70, so the average of each set is the same, equal to 10 (70/7). What we now do is find the sum of the squares of the differences of each number in a set from the average for that set. We now illustrate this with set A:

Set A	7	8	9	10	11	12	13
Difference from average (10)	−3	−2	−1	0	1	2	3
Square of difference	9	4	1	0	1	4	9

We notice here that the square of −3 is the same as the square of +3. The sum of the squares of the differences is 28 and now we find the average square of difference for the seven numbers as:

$$
V = \frac{28}{7} = 4.
$$

This quantity V that describes the spread of the set of numbers is known as the *variance* and taking its square root gives the *standard deviation* of the set of numbers, usually represented by the Greek letter σ (sigma). Here we have:

$$
\sigma_A = \sqrt{V} = \sqrt{4} = 2.
$$

The same process carried out for set B gives:

Set B	1	6	9	10	11	14	19
Difference from average (10)	−9	−4	−1	0	1	4	9
Square of difference	81	16	1	0	1	16	81

The variance is

$$
V = (81 + 16 + 1 + 0 + 1 + 16 + 81)/7 = 196/7 = 28
$$

and hence:

$$\sigma_B = \sqrt{V} = \sqrt{28} = 5.29.$$

The values of the standard deviations, σ_A and σ_B, are measures of the spread of the two sets of numbers, and clearly indicate that set B has a much larger spread than does set A.

Now a normal distribution may come from a large number, perhaps many millions, of quantities and for these quantities, and hence for the distribution function they form, there will be a standard deviation. In fact, the three distributions shown in Fig. 10.5 have variances of 1, 2 and 3 and hence standard deviations of 1, $\sqrt{2}$ and $\sqrt{3}$, respectively.

10.4. Properties of Normal Distributions

An important characteristic of normal distributions is that, essentially, they all have the same shape. You may look at Fig. 10.5 and wonder how that can be so — surely the three curves shown have different shapes! To understand what is meant in saying that all normal curves essentially have the same shape, we consider Fig. 10.6

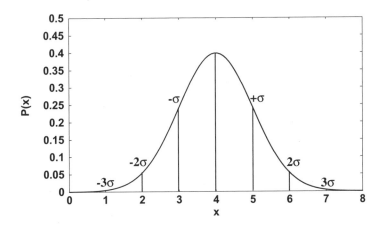

Fig. 10.6. A normal curve with lines at $\pm\sigma$, $\pm2\sigma$ and ±3.

that shows a normal curve with distances $1\sigma, 2\sigma$ and 3σ from the mean, on both sides of the mean, indicated by vertical lines. The total area under the curve is 1: a necessary condition for a sensible distribution curve. The characteristic of all normal distributions, no matter what are their standard deviations, is that:

the area between $-\sigma$ and $+\sigma$ is 0.6826,
the area between -2σ and $+2\sigma$ is 0.9545,
the area between -3σ and $+3\sigma$ is 0.9973.

We now discuss what this means. If a particular quantity has values with a normal distribution, then the probability that one of the values chosen at random differs in magnitude from the mean, either in the positive or negative direction, by 1σ or less, is 0.6826. It follows from this that, since the total area under the curve is 1, we can say that the probability that the quantity differs in magnitude from the mean by more than 1σ is $1 - 0.6826 = 0.3174$, roughly one third. By the same reasoning, we can also see that the probability of having a quantity more than 3σ from the mean is only 0.0027, about 1 chance in 400. In the language of statisticians, a part of the curve beyond a particular specified limit is referred to as a tail of the distribution.

Finding the areas from the mean either out to, or beyond, a particular number of standard deviations for a normal curve is frequently required in statistical work. Table 10.1 gives the area from the mean out to four standard deviations (just in one direction) in steps of 0.1σ. To illustrate the use of this table, if we wish to find the area from the mean out to 1.7 standard deviations from the mean, then we go down the column headed 1.0, down to the row labelled 0.7 and read off the area 0.4554.

The distribution of men's heights shown in Fig. 10.1 has a mean of 1.75 m and a standard deviation of 0.06 m (6 cm). Thus the probability that a man chosen at random is taller than 1.90 m (equivalent to 6' $2\frac{3}{4}$") can be found by noting that 1.90 m is 15 centimetres, or 2.5 σ, from the mean. We can now find the area under the curve from the mean out to 2.5 σ from the sixth entry down in the column headed

Table 10.1. Area from the mean to n standard deviations from the mean for a normal distribution.

n	0.0	1.0	2.0	3.0	4.0
0.0	0.0000	0.3413	0.47725	0.49865	0.49997
0.1	0.0398	0.3643	0.48214	0.49903	
0.2	0.0793	0.3849	0.48610	0.49931	
0.3	0.1179	0.4032	0.48928	0.49952	
0.4	0.1554	0.4192	0.49180	0.49966	
0.5	0.1915	0.4332	0.49379	0.49977	
0.6	0.2257	0.4452	0.49534	0.49984	
0.7	0.2580	0.4554	0.49653	0.49989	
0.8	0.2881	0.4641	0.49744	0.49993	
0.9	0.3159	0.4713	0.49813	0.49995	

2.0 in Table 10.1. This is 0.49379. The total area in one-half of the curve is 0.5, so that the area beyond $2.5\,\sigma$ is $0.5 - 0.49379 = 0.00621$, which is the probability that a man chosen at random has a height greater then 1.90 m. To see what this means, we consider a town with 30,000 adult males. Of these $30,000 \times 0.00621 = 186$ of them should be taller than 1.90 m, equivalent to 6′ $2\frac{3}{4}$″. In the other direction, the height 1.60 m (5′ 3″) is $2.5\,\sigma$ less than the mean height, so 186 of the townsmen should be shorter than this height.

10.5. A Little Necessary Mathematics

10.5.1. *Some special numbers*

A number that occurs a great deal in mathematics is the exponential e given by:

$$e = 2.718281828.$$

It seems a fairly arbitrary number, but it crops up over and over again in mathematics and the sciences in a very natural way. It is not easy to explain to a non-mathematician how and why it occurs so naturally, but it does. One way of defining e is to take the equation:

$$\varepsilon = \left(1 + \frac{1}{n}\right)^n \tag{10.1}$$

and to make n very large — in fact infinity. Here we show the value of ε, starting with $n = 2$ and increasing n to the final value of 1,000,000:

n		ε
2	$(1 + 0.5)^2$	$= 2.250000$
5	$(1 + 0.2)^5$	$= 2.488320$
10	$(1 + 0.1)^{10}$	$= 2.593742$
20	$(1 + 0.05)^{20}$	$= 2.653298$
50	$(1 + 0.02)^{50}$	$= 2.691588$
100	$(1 + 0.01)^{100}$	$= 2.704814$
200	$(1 + 0.005)^{200}$	$= 2.711517$
500	$(1 + 0.002)^{500}$	$= 2.715569$
1,000	$(1 + 0.001)^{1000}$	$= 2.716924$
10,000	$(1 + 0.0001)^{10,000}$	$= 2.718146$
1,000,000	$(1 + 0.000001)^{1,000,000}$	$= 2.718280$

As we make n larger and larger, so ε gets closer and closer to e.

While e cannot be justified as a special number very easily, the same is not true for another frequently occurring number, represented by the Greek letter π (pi), given by:

$$\pi = 3.141592654,$$

which is easily described as the circumference of a circle divided by its diameter. This was used in the simple random number generator described in Equation (8.1).

10.5.2. *Powers of numbers*

Another piece of mathematics that is familiar consists of taking powers, a process that was used in the calculations using Eq. (10.1) to find the successive approximations to e. Thus, for example, we know that the square of 2 is 4 and the cube of 2 is 8, which we can write mathematically as:

$$2^2 = 2 \times 2 = 4 \quad \text{and} \quad 2^3 = 2 \times 2 \times 2 = 8.$$

We see there the notation 2^2 for the square of 2 and 2^3 for the cube of 2. However, mathematically, we can have $2^{2.3}$ and this cannot be written as 2 multiplied by itself a finite number of times. The first step in understanding what this means is to consider the following:

$$2^2 \times 2^3 = (2 \times 2) \times (2 \times 2 \times 2) = 32 = 2^5$$

or, in general,

$$x^a \times x^b = x^{a+b}. \tag{10.2}$$

The general equation can be related to the original specific example by taking $x = 2, a = 2$ and $b = 3$. In fact, this idea can be extended to the product of many powers of x so that, for example, we come to our first rule for handling products of powers:

Rule 1 $\qquad\qquad x^a \times x^b \times x^c \times x^d = x^{a+b+c+d}. \tag{10.3}$

Now we consider what happens when we take a power of a power of a number. The following example, taking the cube of 2 squared, illustrates this:

$$(2^2)^3 = (2 \times 2) \times (2 \times 2) \times (2 \times 2) = 2^{2\times3} = 2^6.$$

This generalizes to the second rule:

Rule 2 $\qquad\qquad (x^a)^b = x^{a\times b}. \tag{10.4}$

We can now understand what is meant by fractional powers of numbers. For example, from Rule 1:

$$2^{1/3} \times 2^{1/3} \times 2^{1/3} = 2^1 = 2,$$

so that $2^{\frac{1}{3}}$ is the cube root of 2, i.e., the number such that the product of three of them gives 2. By generalization, we can see that:

$$x^{1/n} \quad \text{is the } n^{\text{th}} \text{ root of } x, \tag{10.5}$$

i.e., the number such that the product of n of them gives x.

With the above information we are now in a position to understand what $2^{2.3}$ means. According to Rule 2, we can write it as:

$$2^{2.3} = (2^{23})^{\frac{1}{10}},$$

which is the tenth-root of 2^{23}. In practice, it can be found with a scientific hand calculator in the form $2^{2.3}$ and is 4.9246.

We finish with a last little oddity. Using Rule 1, we find:

$$x^a \times x^0 = x^{a+0} = x^a,$$

from which we come to the conclusion that $x^0 = 1$. This is absolutely true: any finite number raised to the power zero equals 1!

What we have established is that any positive quantity can be raised to any power, and that this power can be fractional or even zero. We do not have to worry about interpreting such calculations when we need them: they can easily be found by calculators or computers.

10.6.   The Form of the Normal Distribution

The formula describing the normal distribution involves the exponential e raised to some power: something of the form e^x. However, x is itself a complicated function and there is obviously a problem in showing e^x clearly, if x is a bulky function to express. To help with this problem there is a convention by which one can write:

$$e^x = \exp(x). \tag{10.6}$$

With this notation we can represent the normal (Gaussian) distribution for a quantity x that has a mean value $\bar{x}$ and a standard deviation σ as:

$$P(x) = \frac{1}{\sqrt{2\pi\sigma^2}} \exp\left\{-\frac{(x - \bar{x})^2}{2\sigma^2}\right\}. \tag{10.7}$$

With x representing height, with mean height $\bar{x} = 1.75$ m and with standard deviation $\sigma = 0.06$ m, this produces the curve shown in Fig. 10.1. The factor before "exp" ensures that the curve is properly scaled, so that the area under the curve is 1.

10.7. Random and Systematic Errors

When a scientist carries out an experiment, say, to find the speed of light, she will make various measurements and then combine them to get the quantity she wants. For example, measuring the speed of light may involve a measurement of distance and a measurement of time, and both these measurements would have some associated error. The error may be large, small or even minute, but no matter how much care the scientist takes there will be some error. The errors in the different measurements — and in some experiments there can be many of them — will combine to give an error in the estimated value of the quantity of interest. Now we imagine that the scientist repeats the experiment many times (she would normally not do so in practice) and ends up with a large number of estimates. The distribution of the values of these estimates would be Gaussian, with a mean close to the true value of the quantity and a standard deviation that would depend on the precision of the individual measurements. If the measurements were all made very precisely, then the standard deviation of the distribution would be small; if they individually had large errors, then the standard deviation would be large.

A kind of procedure that involves errors is the measurement of a baseline in surveying. To create a map it is necessary to measure a baseline, perhaps several kilometres long, that will create the correct scale for the mapping procedure. A classical way of measuring a baseline was with a steel tape. The length of the tape was marked off repeatedly over the whole span of the baseline, with each length indicated by score lines on a metal plate attached firmly to the ground. The tape was suspended above the ground, attached to supports adjusted to be at the same level at each end. Because of its weight, the tape was not straight but bowed downwards in a shape known as a catenary, and a correction had to be made for this effect. Considering the complication of this process, the measurements were made with surprising precision. Even earlier, a standard chain 100 feet long was used, which was supported horizontally by wooden coffers each 20 feet long. These were used in the great survey of India, carried out by the British Army at the beginning of the nineteenth century, and completed by Colonel George Everest, whose name was given to the world's highest mountain. Baselines of length 7 to 10 miles were measured, with an accuracy of about one inch: approximately two parts in a million. Every measurement of a single chain length, of which there would be several hundred along the baseline, would involve a small error, sometimes positive and sometimes negative. The sum of these errors gave the total error in the measurement of the baseline and, clearly, positive and negative errors would compensate to some extent. Nevertheless, if the baselines had been measured many times, the estimates would have had a normal distribution centred on a value close to the correct value and with a standard deviation of the order of one inch.

In the above discussion of errors we have assumed that the errors in the measurements were random and equally likely to be negative as positive. However, there is another kind of error for which this supposition is not true. Let us suppose that the steel tape, or the chain, used for the baseline measurement had stretched, which sometimes they had. Now a distance estimated as 100 feet, the assumed length of the chain, is actually slightly longer, so the total distance is going to be underestimated, independent of any random errors. The total baseline measured with a stretched chain, if measured many times, would give a normal distribution of estimated distances, with a standard deviation dependent on the magnitude of the random errors but with a mean shorter than the true distance. This error of the mean is called a systematic error.

Assuming that only random errors are present, if the quantity is measured as X with an estimated standard deviation (usually called a standard error in this context) of σ, then the probability that the true value is in the range $X + σ$ and $X - σ$ is 0.6826 and in the range $X + 2σ$ and $X - 2σ$ is: 0.9545, just as indicated by a normal distribution.

10.8.   Some Examples of the Normal Distribution

10.8.1.   *Electric light bulbs*

Hot filament electric light bulbs are a common and cheap commodity, produced in large numbers by mass-production processes. Inevitably, there is some variation in their characteristics, and the one that most people are concerned with is their lifetime. A more expensive light bulb that lasts longer may be preferred to a cheaper one with a shorter lifetime, no matter what the relative cost per hour, if the effort of changing a bulb is a primary consideration.

We now consider a particular brand of light bulb that is found to have a mean life of 1,000 h and a standard deviation of 100 h. Assuming that the lifetimes have a normal distribution, then what is the probability that a particular bulb will have a lifetime of more than 1,200 h? Since 1,200 h is 2σ from the mean, we are interested in the area of that part of the normal distribution that is more than 2σ from the mean in the positive direction. From Table 10.1, the area between the mean and 2σ from the mean is 0.47725, so the area in the tail more than 2σ from the mean in the positive direction is $0.5 - 0.47725 = 0.02275$. Hence the proportion of bulbs with a lifetime of more than 1,200 h is about 1 in 45. We should also note that this is also the probability of a lifetime of less than 800 h.

If we bought a light bulb and it only lasted 700 hours would we be justified in complaining? Not really: in the sale of millions of light bulbs some will inevitably fail early. We would not be rushing to pay the shop extra money if we had a bulb of unusually long lifetime. Again, the pattern of use must be taken into account, since constantly turning a bulb on and off subjects it to thermal shocks that shorten its lifetime. In the museum in Fort Myers, Florida, devoted to the work of the inventor Thomas Edison (1847–1913), who invented the first practical filament light bulb in 1879, one of the original bulbs is still burning. It has never been switched off.

10.8.2. *People on trolleys and under-used resources*

Occasionally, a story hits the news media in the United Kingdom concerning an individual left on a hospital trolley in a hospital corridor for several hours. Clearly, the National Health Service is under-resourced, inadequate and failing, or so the tabloid press would have you believe!

The provision of a health service is made complicated by the fact that the demands on it are spasmodic and unpredictable. A cold snap or an influenza epidemic may mean a sudden large increase in the admission of elderly patients. It could be argued that, although the times of occurrence of these eventualities are unpredictable, it is known that they occur from time to time, so provision should be made for them. Let us see what the implications of that argument are by considering a hypothetical hospital with 1,000 beds. The demand for beds, estimated over a long period of time, suggests that the average demand on any day is 900 beds with a standard deviation of 50. Now, on the (dubious) assumption that the daily demand follows a normal distribution, we consider the following questions.

(i) On how many days per year, on average, is the hospital unable to provide all the beds required?

If more than 1,000 beds were required, then the demand would be more than 2σ from the average. From the light bulb example we know that the area of a single tail more than 2σ from the mean is 0.02275, so the number of days per year that demand cannot be met is $365 \times 0.02275 = 8$ days to the nearest whole number. Sometimes the excess demand can be moved to a neighbouring hospital, but if all local hospitals are under pressure then the hospital trolley expedient becomes necessary.

(ii) On how many days per year, on average, is the hospital working at less than 90% capacity?

Since 90% capacity is 900 beds (the average demand), then clearly on 50% of days, or 188 days per year, the hospital resources are being underused. Since, under any reasonable employment regime, staff cannot be engaged and dismissed to match demands fluctuating on a timescale of a few days, some inefficiency must be present in the system.

This is the problem of providing any resource for which there is a fluctuating demand. If the demand were constant, then the resource could be made to exactly match the demand, every demand would be met, and the resource could be provided in the most economical way. This is all a question of priorities. By increasing the number of beds to 1,050, the incidence of being unable to meet demand falls to 1 day per two years. On the other hand, there are then 246 days per year of underused resources, including staff. Managers have various ways of trying to optimize the efficiency of their hospitals: moving patients between hospitals to try to balance over-demand in one hospital by spare capacity in another has already been mentioned. Employing agency staff on a short-term basis as needed is another possibility, but agency staff are considerably more expensive than permanent staff.

The actual situation that hospitals have to face will actually be much worse than that described above: demands can increase dramatically due to an emergency, such as an epidemic, a train crash or a terrorist attack. The detailed structure of the hospital provision also plays a role so that, for example, intensive-care provision is expensive and the number of beds devoted to it is limited. In some circumstances, the hospital may have spare beds, but not the right kind of beds.

The normal distribution, that we have used to illustrate the hospital capacity problem, will certainly not be an adequate description of the extreme and rapid fluctuations of demand that can occur. The stories of patients on trolleys are certainly worthy of report, and to the person affected, and his or her relatives, the statistics of the situation may not seem relevant. However, such stories are often driven by political agendas; those who are not directly involved, and should therefore be able to take an objective view, should be aware of the problems being faced by hospitals by the very nature of the service they provide. By spending enough money, a system can be set up in which hospitals can meet every demand and, alternatively, by ensuring that excess capacity is kept to a minimum, hospitals can be run economically — but not both at once. Again, it is a question of choice, not of good and bad.

Problems 10

10.1 Find the average, the variance and the standard deviation of the following set of numbers:

$$2 \quad 4 \quad 6 \quad 7 \quad 8 \quad 10 \quad 12$$

10.2 Write in the form 3^n the following:

(i) $3^2 + 3^4 + 3^5$
(ii) $(3^3)^4$
(iii) $(3^{13})^{1/10}$

10.3 The average daily sale of a daily paper is 52,000, with a standard deviation of 2,000. On how many days a year will its sales be:

(i) below 47,000?
(ii) above 55,000?

Chapter 11

Statistics: The Collection and Analysis of Numerical Data

A great multitude, which no man could number (Revelation 7:9)

11.1. Too Much Information

We consider a Minister of Health who wishes to keep a check on the birth weights of babies born in each year, so as to be able to see if there are any significant trends in the country as a whole. Regional health offices record the birth weights of all the babies in their areas and each sends a complete list of weights to the ministry. The Minister is now presented with a document, the size of a large telephone directory, containing about three-quarters of a million numbers. She flicks over the pages and sees the list of weights, 3.228 kg, 2.984 kg, 3.591 kg etc., and she just receives conformation of what she knew anyway: that most babies weigh about 3 kg, or a little more, at birth.

A useful number is the average of the weights and if the information were provided by each region on some electronic medium for computer input, then the average could be found quickly. The average from year to year could then be compared to see what trend, if any, there was. Another quantity of interest would be the spread of the weights, which would best be expressed by the standard deviation, or variance (Section 10.3), of the babies' weights. Again, if the information were provided in electronic form this could readily be found.

Actually, there was no need for the local regional health offices to have sent all the individual weights in the first place. The weights both in the individual regions and in the country as a whole would have a normal distribution. If each region had submitted the number of babies born, their average weight and the standard deviation (or variance) of their weights, then that would be a complete description of the situation in their area. We shall describe the way in which the regional distributions can be combined to give an overall national distribution.

11.2. Another Way of Finding the Variance

In Section 10.3 we described the variance of a set of numbers as: *the average of the square of the difference between each number and the average of the numbers*. While that is a perfectly correct definition of the variance, and the way that we worked it out for the distributions A and B, it does challenge the imagination somewhat. With a little mathematical manipulation the variance can be expressed in an alternative way as: *the average of the square minus the square of the average*. We now apply this way of calculating variance to the sets A and B given in Section 10.3 to check that it gives the same results as we found previously:

Set A	7	8	9	10	11	12	13
Square	49	64	81	100	121	144	169

The average of the square is

$$\overline{x^2} = \frac{49 + 64 + 81 + 100 + 121 + 144 + 169}{7} = 104. \tag{11.1}$$

Notice the terminology: $\overline{x^2}$. The bar over the top means "average of", so $\overline{x^2}$ means the average of x^2 where the individual numbers of set A are the x's. We also found previously that the average of set A, that we indicate as $\bar{x}$, is 10, so the variance is:

$$V = \overline{x^2} - \overline{x}^2 = 104 - 10^2 = 4, \tag{11.2}$$

the same as we found previously. Similarly, for set B:

Set B	1	6	9	10	11	14	19
Square	1	36	81	100	121	196	361

The average of the square is:

$$\overline{x^2} = \frac{1 + 36 + 81 + 100 + 121 + 196 + 361}{7} = 128. \tag{11.3}$$

Again the average is 10, so the variance is:

$$V = \overline{x^2} - \overline{x}^{-2} = 128 - 10^2 = 28, \tag{11.4}$$

as found previously.

We can now use these results to see how to convert the regional birth weight data into that for the nation as a whole.

11.3. From Regional to National Statistics

We now consider one of the regions, designated by the letter j, for which the number of babies born was N_j, the average weight was $\overline{w}_j$ and the variance of the weights was V_j. Since, by our latest definition of variance,

$$V_j = \overline{w_j^2} - \overline{w}_j^2, \tag{11.5}$$

we find by rearranging

$$\overline{w_j^2} = V_j + \overline{w}_j^2. \tag{11.6}$$

That means for each of the regions, since we are given V_j and $\overline{w}_j$, we can find $\overline{w_j^2}$.

We are now in a position to find the average baby weight and the average of the square of the babies' weight for all the babies combined from all the regions. The average baby weight is:

$$\overline{w_{all}} = \frac{\sum_{j=1}^{M} N_j \overline{w_j}}{\sum_{j=1}^{M} N_j}. \tag{11.7}$$

The summation symbol Σ was explained in relation to Equation (3.2). The term $N_j \overline{w_j}$ is the total weight of all the babies born in region j, and summing this over all the regions gives the total weight of all the babies born in that year for the whole nation (about 2,000 tonnes for the United Kingdom!). The divisor in (11.7) is the total number of babies born, so dividing the total weight of the babies by the number of babies gives the average weight of the babies for the whole country.

By an exactly similar process we can find the average *squared weight* of the babies. This is:

$$\overline{w_{all}^2} = \frac{\sum_{j=1}^{M} N_j \overline{w_j^2}}{\sum_{j=1}^{M} N_j}. \tag{11.8}$$

The top of Equation (11.8) is the sum of the squared weights for all the babies in the country and dividing by the total number of babies gives the average squared weight. Now by combining $\overline{w_{all}}$ and $\overline{w_{all}^2}$ we can find the variance for all the babies as:

$$V_{all} = \overline{w_{all}^2} - \overline{w_{all}}^2. \tag{11.9}$$

The Minister has now combined the information from the regions to find the total number of babies born, their mean weight and the variance of their weights — and all that each region had to send in was three numbers. In Table 11.1 we show the results of such an exercise for a hypothetical set of data. The fourth column shows the standard deviation (square root of the variance) of the weights in each region.

The averages, at the bottom of columns three and five, are found from Equations (11.7) and (11.8), respectively.

Table 11.1. Data for the birth weights of babies born in different regions.

Region	N	$\bar{w}$ (kg)	σ (kg)	$\overline{w^2}$ (kg²)
South	66,296	3.062	0.251	9.438845
South-east	108,515	2.997	0.239	9.039130
South-west	64,621	3.185	0.267	10.215514
West	76,225	3.002	0.224	9.062180
Central	93,496	2.996	0.250	9.038516
East	41,337	3.099	0.231	9.657162
North-east	104,212	3.101	0.237	9.672370
North-west	82,358	3.011	0.226	9.117197
Far-north	37,362	3.167	0.219	10.077850
Outer Isles	26,219	3.178	0.220	10.148084
Average (sum)	(700,641)	3.05987		9.424592

The table directly gives the total number of babies born, 700,641, and their average weight, 3.05987 kg. The variance of the weights is:

$$V = \overline{w^2_{all}} - \overline{w_{all}}^2 = 9.424592 - (3.05987)^2 = 0.0617875 \text{ kg}^2,$$

corresponding to a standard deviation:

$$\sigma = \sqrt{V} = 0.2486 \text{ kg}.$$

The Minister now has available all the information she needs.

The United Kingdom Office for National Statistics collects and analyses statistical data over a wide range of national activities concerned with health, the economy, demography, employment and business. Many of its analyses are based on the assumption of a normal distribution but there are other kinds of distribution that we shall now consider.

Problems 11

11.1 Use Equation (11.2) to find the variance of the set of numbers given in Problem 10.1.

11.2 Four schools in a particular district send in the following information concerning the heights of their 11-year-old boy pupils:

School	Number of pupils	Mean height	Standard deviation
1	62	1.352 m	0.091 m
2	47	1.267 m	0.086 m
3	54	1.411 m	0.089 m
4	50	1.372 m	0.090 m

What is the mean height and standard deviation of heights of 11-year-old boys in the district?

Chapter 12

The Poisson Distribution and Death by Horse Kicks

Life is good for only two things, discovering mathematics and teaching mathematics (Siméon-Denis Poisson (1781–1840), *Mathematics Magazine*, 1891)

12.1. Rare Events

Although the normal distribution is by far the most important naturally-occurring distribution, there are others, and one of these, of considerable importance, we now describe.

Siméon-Denis Poisson (1781–1840), pictured in Fig. 12.1, was a French mathematician and physicist whose contributions covered an enormous range of topics and who must be considered one of the scientific giants of the nineteenth century. In 1837 he wrote a paper with a title that translates into English as *Research on the Probability of Criminal and Civil Verdicts*, in which he developed and described what is now known as the *Poisson distribution*. Although it never again featured in any of his mathematical publications, this distribution turns out to be of great importance in dealing with the statistics of many real-life situations.

To illustrate an application of the Poisson distribution, let us consider the problem of taking a census of traffic flow on a quiet country road. We decide to divide time up into one minute intervals and in each interval the number of cars passing is recorded.

Fig. 12.1. Siméon-Denis Poisson, lithography by François-Seraphin Delpech.

In one hour, the 60 numbers recorded are:

$$
\begin{array}{cccccccccccc}
0 & 0 & 1 & 0 & 1 & 2 & 1 & 0 & 0 & 1 & 1 & 0 \\
1 & 1 & 0 & 2 & 0 & 0 & 0 & 1 & 3 & 1 & 0 & 1 \\
0 & 0 & 0 & 1 & 1 & 2 & 1 & 0 & 1 & 1 & 0 & 0 \\
0 & 0 & 1 & 2 & 1 & 1 & 0 & 0 & 0 & 0 & 0 & 3 \\
0 & 1 & 1 & 2 & 0 & 0 & 1 & 0 & 1 & 2 & 0 & 1 \\
\end{array}
$$

These results can be represented by the histogram, Fig. 12.2, which shows the number of intervals for each number of cars passing.

The numbers of cars per one-minute interval were small: for many intervals, none, for somewhat fewer intervals, one, for fewer intervals, two, and for even fewer intervals, three. It is to explain the variation in frequency in situations like this that Poisson developed the theory of what is now called the *Poisson distribution*. We can illustrate the mathematical nature of this distribution by taking another situation, where the number of trials (minute intervals in the case we have just taken) is large but the numbers of events per trial (the passing of cars) are small.

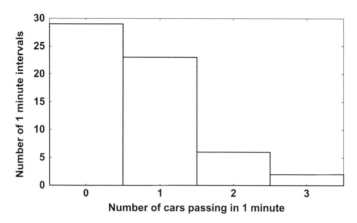

Fig. 12.2.   Histogram for numbers of cars passing per minute on a quiet country road.

12.2.   Typing a Manuscript

We consider the task of typing a manuscript of 100 pages. Being human, and hence prone to error, the typist makes occasional typing errors although, for a good typist, the average number of errors per page would be small. When the pages are proof-read, the number of errors per page is found, with the following results:

Number of errors	0	1	2	3	4	5	more than 5
Number of pages	30	36	22	9	2	1	0

This distribution is illustrated as a block diagram in Fig. 12.3.

The first thing we can find is the average number of errors per page. The total number of errors is:

$$(30 \times 0) + (36 \times 1) + (22 \times 2) + (9 \times 3) + (2 \times 4) + (1 \times 5) = 120$$

and, hence, the average number of errors per page is:

$$a = \frac{120}{100} = 1.2.$$

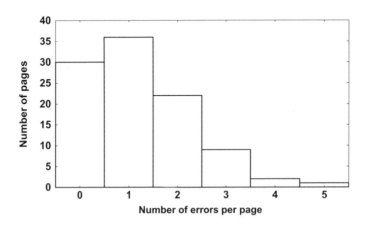

Fig. 12.3.   The numbers of pages with particular numbers of typing errors.

The Poisson distribution is completely defined in terms of the average, a. Starting with the number of pages with no errors, we multiply that by a to obtain:

$$30 \times a = 36,$$

which is the number of pages with one error.

Now we multiply the number of pages with one error by a and divide by 2:

$$36 \times a \div 2 = 21.6.$$

The nearest whole number is 22, the number of pages with two errors.

Now we multiply the number of pages with two errors by a and divide by 3:

$$22 \times a \div 3 = 8.8.$$

The nearest whole number is 9, the number of pages with three errors.

Now we multiply the number of pages with three errors by *a* and divide by 4:

$$9 \times a \div 4 = 2.7.$$

The nearest whole number is 3, but the actual number of pages with four errors is 2. The calculated and observed numbers of errors do not exactly agree in this case, which is partly due to taking nearest whole numbers in the calculated values.

Now we multiply the number of pages with four errors by *a* and divide by 5:

$$3 \times a \div 5 = 0.7.$$

The nearest whole number is 1, the number of pages with five errors.

Although this process gives the numbers of pages with different numbers of errors related by a sequence of factors, it is better to have an explicit formula to give the number of pages with a specified number of errors.

12.3.   The Poisson Distribution as a Formula

Expressing a Poisson distribution in a mathematical form, the fraction of cases with *r* events (error on a page, cars passing in one minute), where the average is *a* is:

$$F(r) = \frac{e^{-a}a^r}{r!}. \tag{12.1}$$

The two terms at the top of this formula have been explained in Section 10.5. The term at the bottom is factorial *r* that was described in Section 4.1.

Let us see how the formula works for the typist's errors, where the average number per page is 1.2. With a calculator, we find that: $e^{-a} = e^{-1.2} = 0.3012$. The fraction of pages with zero errors is thus:

$$F(0) = \frac{e^{-1.2} \times 1.2^0}{0!} = 0.3012$$

since, as we have previously seen: $1.2^0 = 1$ and $0! = 1$. Because there are 100 pages in all, to the nearest whole number, 30 of them will be error-free.

The fraction of pages with one error is:

$$F(1) = \frac{e^{-1.2} \times 1.2}{1!} = 0.3614, \quad \text{giving 36 pages with one error.}$$

Similarly,

$$F(2) = \frac{e^{-1.2} \times 1.2^2}{2!} = 0.2169, \quad \text{giving 22 pages with two errors.}$$

$$F(3) = \frac{e^{-1.2} \times 1.2^3}{3!} = 0.0867, \quad \text{giving 9 pages with three errors.}$$

$$F(4) = \frac{e^{-1.2} \times 1.2^4}{4!} = 0.0260, \quad \text{giving 3 pages with four errors.}$$

$$F(5) = \frac{e^{-1.2} \times 1.2^5}{5!} = 0.0062, \quad \text{giving 1 page with five errors.}$$

Once again, there is a discrepancy for the number of pages with four errors, but this is due to the numbers being rounded-off — there cannot be 36.14 pages with one error.

Like any sensible distribution giving probabilities, the sum of the probabilities for all the possible outcomes must be 1, that is to say, that it is completely certain that there must be some outcome. Thus we can write:

$$F(0) + F(1) + F(2) + F(3) + F(4) + F(5) + F(6) + \cdots = 1 \quad (12.2)$$

where, on the left-hand side, we have an infinite number of terms. The sum of the probabilities for $F(0)$ to $F(5)$ for the typing errors is 0.9984, since the probabilities from $F(6)$ onwards are missing.

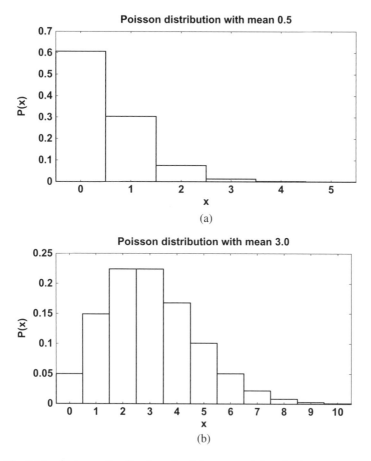

Fig. 12.4. Poisson distributions for (a) average 0.5 and (b) average 3.0.

To get a feeling for Poisson distributions, we now show as histograms in Fig. 12.4 the distributions for averages 0.5 and 3.0.

12.4. Death by Horse Kicks

One of the most famous applications of the Poisson distribution was made by a Russian statistician, Ladislaus Bortkiewicz (1868–1931), pictured in Fig. 12.5.

At the end of the nineteenth century cavalry units formed a component of most armies, and once in a while someone in a cavalry

Fig. 12.5. Ladislaus Bortkiewicz.

(From *Reichshandbuch der Deutschen Gesellschaft, Bd, I: Das Handbuch der Pers nlichkeiten in Wort und Bild*, 1930, S. 188)

Table 12.1. Statistics of deaths by horse kicks for 200 cavalry unit years.

Number of deaths per unit per year	Number of unit years
0	109
1	65
2	22
3	3
4	1
More than 4	0

unit would be killed by a horse kick. Bortkiewicz took the record of such deaths for 10 cavalry units of the Prussian Army over a period of 20 years, from 1875 to 1894, and analysed them statistically. For the 200 cavalry unit years, the information he derived is shown in Table 12.1.

Bortkiewicz showed that the number of deaths per cavalry unit year followed a Poisson distribution. From his figures, the total number of deaths was:

$$65 \times 1 + 22 \times 2 + 3 \times 3 + 1 \times 4 = 122,$$

so the average number of deaths per cavalry unit year was $\frac{122}{200} = 0.61$.

Equation (12.1) gives the proportion of cases with r events (deaths), so the number of cavalry unit years in which no deaths would be expected, given the average, is:

$$N(0) = 200 \times \frac{e^{-0.61} \times 0.61^0}{0!} = 108.7,$$

or 109, rounded off to the nearest whole number. This is exactly the figure found in Table 12.1. Similarly, with the rounded-off figure in brackets:

$$N(1) = 200 \times \frac{e^{-0.61} \times 0.61^1}{1!} = 66.3 \ (66),$$

$$N(2) = 200 \times \frac{e^{-0.61} \times 0.61^2}{2!} = 20.2 (20),$$

$$N(3) = 200 \times \frac{e^{-0.61} \times 0.61^3}{3!} = 4.0(4),$$

$$N(4) = 200 \times \frac{e^{-0.61} \times 0.61^4}{4!} = 0.6 \ (1),$$

and the similarity of these figures to those given in Table 12.1 will be evident. Indeed, a χ^2 test shows no significant difference between the expected numbers we have just calculated and the observed numbers found by Bortkiewicz.

The Poisson distribution is important in many areas of life and we now give two further examples.

12.5. Some Other Examples of the Poisson Distribution

12.5.1. *Flying Bomb Attacks on London*

During the final period of the Second World War the Germans deployed a weapon, mainly for bombarding London and the south-east of England, known as the *flying bomb*. It was a small pilot-less aircraft, driven by a simple ramjet motor that travelled at 575 km hour^{-1} and had a range of 240 km. These were launched from a ramp pointing towards London and, when the fuel ran out, they fell to the ground as a bomb containing one tonne of explosive. Of the 8,600 that were launched, roughly one-half penetrated the defences, the rest being shot down by anti-aircraft fire and fighter planes.

Clearly, they were a very imprecise weapon and depended for their success on the large size of the target, which could hardly be missed. They were at the mercy of unpredictable tail winds, head winds and side winds, so their falls were distributed all over the London area. There was a suspicion that the fall sites of the bombs tended to occur in clusters, so it was decided to apply a statistical test on the fall of the bombs in London up to that time. London was divided up into 576 squares with sides 0.5 km and the number of bombs that had fallen in each square was recorded. The results are shown in Table 12.2.

The total number of bombs falling in this period was:

$$211 \times 1 + 93 \times 2 + 35 \times 3 + 7 \times 4 + 1 \times 5 = 535,$$

so the average number of bombs per square equalled $\frac{535}{576} = 0.9288$. The final column of Table 12.2 shows the numbers expected from the

Table 12.2. The number flying bombs landing within $\frac{1}{4}$ km^2 squares in London.[a]

Number of bombs	Number of squares	Poisson distribution
0	229	227.5
1	211	211.3
2	93	98.1
3	35	30.4
4	7	7.1
5	1	1.3

[a]R.D. Clark, 1946, "An Application of the Poisson Distribution", *Journal of the Institute of Actuaries* **72**, 481.

Poisson distribution. The similarity between the observations and the Poisson distribution is evident, and a χ^2 test shows that there is no significant difference between the two. The idea of clustering was not supported by this analysis.

12.5.2.   *Clustering of a disease*

The medical authorities are often interested in what are known as *clusters* of particular diseases. If the incidence of a condition, say, testicular cancer, is higher than would be expected in a particular region, then one might look for contributory causes of a local nature, such as particular industrial processes or the use of pesticides. Let us consider a hypothetical example.

A condition affecting young men is known to occur with a frequency of 2.2 per 100,000 men. In a city with 50,000 young men, 8 are found to have the condition. Is there any evidence that this is a significant cluster of incidence?

The expected number for 50,000 young men is 1.1, so we are clearly interested in a Poisson distribution with that average. In testing for the significance of an outcome greater than the expectation, we find the total probabilities of *the actual result and all results which are more extreme*, i.e., the combined probabilities of 8, 9, 10, 11 etc., up to infinity. Since from (12.2) the sum of all the probabilities of a Poisson distribution must equal unity, we can find the probability of 8 or more cases by taking away from unity the sum of the probabilities of 7 or less cases. This gives:

$$P(\geq 8) = 1 - F(0) - F(1) - F(2) - F(3) - F(4) - F(5) - F(6) - F(7)$$

or

$$P(\geq 8) = 1 - e^{-1.1}$$
$$\times \left(1 + 1.1 + \frac{1.1^2}{2!} + \frac{1.1^3}{3!} + \frac{1.1^4}{4!} + \frac{1.1^5}{5!} + \frac{1.1^6}{6!} + \frac{1.1^7}{7!}\right)$$
$$= 0.000020124.$$

Now this is a very small probability, so we may conclude that this is a significant cluster and that some local factor is operating to produce the condition. This may be genetic, cultural, dietary or of some other origin and it would take further investigation to find out which.

In judging this significance we have to beware to not jump to conclusions. Suppose the answer had been 0.01, i.e., 1 chance in 100 — could we conclude then that the cluster was significant? Let us say that there are 5,000,000 young men in the region that gave the incidence rate of 2.2 per 100,000. Then there are 100 sets of 50,000 men so, on average, one of those sets of 50,000 young men would give the outcome we found. This would come about just by the vagaries of a random distribution of cases, so one could not then conclude that there was a significant cluster. Statistics must be treated with care: they can be misleading but also, sometimes, those with bad intent can attempt to use them to mislead!

12.5.3. *Some further examples*

To finish our description of the Poisson distribution, we give here a sample list of other situations in which the distribution plays a role.

(i) The number of calls at a call centre per minute. By having too few operators the customer may have a long wait, leading to excessive customer dissatisfaction. At the other extreme, having a sufficient number of operators to meet the greatest peak demand will mean that operators are often idle, leading to inefficiency in the running of the call centre. By analysing the demand, which will be in terms of Poisson statistics, a compromise number of operators can be found that will be acceptably economic while, at the same time, giving an acceptable service to the customer.

(ii) The number of light bulbs that fail per day in a large building. One aspect of servicing a large business premises is to replace failing light bulbs. There might be an average of seven failing per day, but the actual number per day, taken over a long period of time, will give a Poisson distribution with average seven.

(iii) The number of mutations in a given stretch of DNA for a given dose of radiation. DNA is the material within the animal cell that controls the genetic make-up of the individual. It consists of a long chain of many thousands of units of which there are four basic types. When exposed to radiation, DNA can become damaged and will then reconstruct itself, sometimes but not always giving modification of the original genetic information it contains — i.e., a mutant has been created. The number of mutations for a given radiation dose will be a matter of chance but, in a statistical sense, the numbers will follow a Poisson distribution.

Problems 12

12.1 The number of faulty components in a manufacturing process follows a Poisson distribution with a mean of 1 per 100 components. What is the probability that a random selection of 100 components will have:

    (i) no faults?
    (ii) 1 fault?
    (iii) 3 faults?
    (iv) 5 faults?

12.2 Large asteroids fall onto Earth at the rate of 1 per 10,000,000 years. What is the probability that at least 1 will fall on Earth within the next 1,000,000 years?

12.3 The average number of electric light bulbs failing per day in a factory is 10. On how many days per year would the number failing be 15?

Chapter 13

Predicting Voting Patterns

Die Politik ist keine exakte Wissenschaft (Politics is not an exact science) (Otto von Bismarck, 18th December 1863, Speech to the Prussian Chamber)

13.1. Election Polls

In democratic countries there is always a natural interest in knowing the support given to the various parties that, from time to time, contest national and local elections. Ideally, one would like to know the voting intention of everyone who intends to vote, but this is tantamount to actually holding an election. As an alternative, polling organizations collect the opinions of *samples* of the population, and then give estimates of the voting intentions of the whole population based on the declared intentions of those sampled.

To obtain a good sample is a complicated exercise. To start with, one must be sure that those sampled are a good representative selection of the population as a whole. There are various ways of categorizing members of the population, for example, by social class, age, gender, income, ethnicity, religion or region of the country. If the sample is taken just from those entering an expensive jewellery store in the West End of London, the results of the poll would be as unlikely to be representative of the whole population as sampling just those entering a fish and chip shop in a run-down northern inner-city area. The clientele of both those establishments should be part of the polled sample, but not its totality. The major polling organizations show great skill in getting good samples and in general their results are now reliable. They have learned from many early mistakes. In the 1948 presidential election in the United States, all the polls showed that the Republican, Thomas E. Dewey, had a comfortable lead of between 5% and 15% over his Democratic opponent, Harry S. Truman. The Chicago Daily Tribune newspaper, pressured by a printing deadline before the actual result was known, brought out an edition with a huge headline "DEWEY DEFEATS TRUMAN". In the event, Truman won the election with a 4.4% advantage over his opponent. It turned out that the sampling of public opinion had been carried out by telephone; Dewey certainly had majority support from those that owned telephones in 1948, but hardly any support from the many without a telephone at that time.

For the purposes of our discussion of polling, we shall assume that the sample of the population is a good representative one and, initially, that the choice is basically between two alternative parties. Later, we shall consider the situation when there are three or more choices of party or individual.

13.2.   Polling Statistics

We consider a situation where just two parties, the National Democrats and the People's Party, are offering themselves to the electorate. A poll of 1,000 representative members of the public is taken and indicates their support as follows:

National Democrats   522   People's Party   478

Translated into proportions, the voting intentions of the sample become:

National Democrats   0.522   People's Party   0.478

On the face of it, the National Democrats party has a clear lead and must be considered as likely to win a majority of the popular vote. However, can it be said that they are *certain* to do so? Clearly not: the possibility must exist that there is actually a majority in favour of the People's Party but that, just by chance, the poll sample included a majority of National Democrats supporters. The reality of the situation is that there is a finite, but small, probability that the People's Party will win and we now see how to estimate this mathematically.

Let the size of the poll be N, in this case 1,000, and the proportion of the poll sample supporting the National Democrats be p, in this case 0.522. The mathematics tells us that the most likely outcome of the election is that a proportion p will vote for the National Democrats. However, the actual proportion of the electorate supporting the National Democrats could be different from p, say, r. Theory shows that, on the basis of the polling evidence, the unknown value of r has an approximately normal probability distribution with mean:

$$\bar{r} = p \tag{13.1}$$

and with a standard deviation:

$$\sigma_r = \sqrt{\frac{p(1-p)}{N}}. \tag{13.2}$$

Let us see what this means for our hypothetical example. We have from Equation (13.1) $\bar{r} = p = 0.522$ and from Equation (13.2):

$$\sigma_r = \sqrt{\frac{0.522 \times 0.478}{1,000}} = 0.0158.$$

We show this probability distribution in Fig. 13.1. Getting a majority of the popular vote means getting more than 0.5 of the votes and this value is also shown in the figure. The form of the normal distribution, as illustrated in this figure, has unit area under the curve. The probability that the National Democrats will get more of the popular vote is the area to the right of the line indicating $r = 0.5$, and the area to the left is the probability that the People's Party will come out on top. We shall now find what these probabilities are.

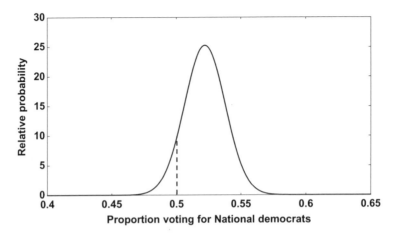

Fig. 13.1.   The probability distribution of those voting for the National Democrats.

The line $r = 0.5$ is:

$$\frac{0.522 - 0.5}{0.0158} = 1.4\sigma_r$$

from the mean of the distribution. From Table 10.1 the area from the mean out to 1.4 standard deviations from the mean is 0.4192. Hence the probability that the National Democrats will get most of the popular vote is the sum of the area to the right of the mean (0.5 because it is one half of the area under the curve), and the area to the left of the mean out to $r = 0.5$ (0.419 which is the area out to $1.4\sigma_p$) giving a total probability 0.919. This indicates that the probability, on the basis of the poll, that the People's Party will win is $1.0 - 0.919 = 0.081$: unlikely but not impossible.

If the analysis had been done from the aspect of looking at the probability distribution of support for the People's Party, then the distribution would have been centred on 0.478 but the standard deviation would have been the same. The line $r = 0.5$ would still be $1.4\sigma_p$ from the mean of the distribution, but in the other direction, and the probability deduced for the likelihood that the People's Party would come out on top would be found as before: 0.081.

We have stressed all the time the concept "proportion of the popular vote" because how that would turn out in terms of seats in the parliament would depend on the electoral system. In a system giving proportional representation, the party with a majority of the popular vote would also have the most seats. In a first-past-the-post system, as prevails in the United Kingdom, this is not necessarily so.

The size of the poll has a considerable effect on the significance of the results. We suppose that the poll size was 2,000 and, again, the proportion of those voting for the National Democrats was 0.522. Using Equation (13.2), the standard deviation now becomes:

$$\sigma_r = \sqrt{\frac{0.522 \times 0.478}{2,000}} = 0.0112.$$

The line $r = 0.5$ is now:

$$\frac{0.522 - 0.5}{0.0112} = 2.0\sigma_r$$

from the mean of the distribution. The area from the mean to 2.0 standard deviations from the mean is, from Table 10.1, 0.47725 and, from this, the probability of the National Democrats getting more of the popular vote is 0.97725. The chance that the People's Party would win has reduced from 0.081, or about 1 chance in 12, to 0.023, or about 1 chance in 43, from the same proportions in a larger poll.

13.3. Combining Polling Samples

There are a number of different polling organizations, differing slightly in their polling techniques but all giving worthwhile estimates of voting intentions. Assuming that they are all equally valid, then it is possible to combine them to give an estimate of voting intentions more precise than any individual one of them. However, care should be taken in making such combinations: for example, party support may vary through a pre-election period, so the data for the separate polls should not be widely separated in time. We consider the following information from three polling organizations in a two-party election:

Poll X poll size 1,000 National Democrats 521 People's Party 479
Poll Y poll size 1,000 National Democrats 517 People's Party 483
Poll Z poll size 1,000 National Democrats 495 People's Party 505
For poll X alone:

$$\sigma_X = \sqrt{\frac{0.521 \times 0.479}{1000}} = 0.0158.$$

Support for the National Democrats is thus $\frac{0.021}{0.0158}\sigma_X = 1.33\sigma_X$ from the level 0.5, beyond which they are the majority party. The area from the mean of the probability distribution (0.521) to the critical 0.5 level of support is, from Table 10.1, equal to 0.408, making the probability of a National Democrats victory 0.908. Similarly, the probability of victory by the National Democrats is 0.860 from poll Y alone and 0.376 from poll Z alone (it predicts a People's Party victory).

When the three polls are combined we find that the indicated fractional support for the National Democrats is 0.511 on a poll of size 3 000. For this combined data:

$$\sigma_{XYZ} = \sqrt{\frac{0.511 \times 0.489}{3000}} = 0.0913$$

and support for the National Democrats is $\frac{0.011}{0.0093}\sigma_{XYZ} = 1.18\sigma_{XYZ}$ from the critical 0.5 level. From Table 10.1 this gives a probability of 0.881 for a National Democrats victory.

Polls taken at different times during an election campaign can be quite informative of trends in the voting intentions of the electorate. However, as with all human activities, there is always some unpredictability, even when the polls are properly conducted.

13.4. Polling with More than Two Parties

Once there are more than two choices of party or individual, the analysis of polls is somewhat more difficult. We now consider a three-party system, with parties A, B and C, for which a poll of size N gave a proportion p_A for party A, p_B for party B and p_c for party C. We have the condition that:

$$p_A + p_B + p_C = 1. \tag{13.3}$$

To find out what the range of possible support is for party A, the simplest approach is to consider a two-choice scenario, where the choice is party A or not-party-A. The probability distribution of the proportion voting for party A can be regarded as a normal distribution, with mean p_A and standard deviation:

$$\sigma_A = \sqrt{\frac{p_A(1 - p_A)}{N}} = \sqrt{\frac{p_A(p_B + p_C)}{N}}. \tag{13.4}$$

Swapping around A, B and C gives the means and standard deviations for the support of the other parties.

Consider a case where a poll of size 2,000 gives $p_A = 0.4$, $p_B = 0.35$ and $p_C = 0.25$. Then the expected means and standard deviations for the parties are:

Party A

$$\bar{p}_A = 0.4 \quad \sigma_A = \sqrt{\frac{0.4 \times 0.6}{2,000}} = 0.0110$$

Party B

$$\bar{p}_B = 0.35 \quad \sigma_B = \sqrt{\frac{0.35 \times 0.65}{2,000}} = 0.0107$$

Party C

$$\bar{p}_C = 0.25 \quad \sigma_C = \sqrt{\frac{0.25 \times 0.75}{2,000}} = 0.0097$$

A precise mathematical analysis of this situation is quite complicated; when there were two parties the probability estimates for getting a majority of the vote were the same, whether one considered the distribution for one party or the other. A gain by one party relative to the poll estimate completely defined the loss by the other and the probabilities of gain and loss were consistent. Here, with three parties, a particular gain by one party involves a net loss by the other two, but the pattern of loss between the other two parties is not defined. However, from the point of view of getting a general impression of where the parties stand, and by how much the actual result is likely to differ from the mean, the individual means and standard deviations as given above serve their purpose well. As an illustration, a poll taken just before the 2005 election in the United Kingdom, with a poll size of 2,000, gave the following percentages, with the actual voting percentage in parentheses:

Labour	37.0%	(35.3%)
Conservative	32.5%	(32.3%)
Liberal Democrat	24.0%	(22.1%)

The minor parties accounted for the remaining few percent. On the basis that votes were either Labour or non-Labour, the standard deviation for the Labour vote was:

$$\sigma_{Lab} = \sqrt{\frac{0.37 \times 0.63}{2,000}} = 0.011,$$

so that the actual vote was:

$$\frac{0.017}{0.011} = 1.55\sigma_{Lab}$$

below the poll prediction. Similarly, the Conservative vote was 0.2 standard deviations below, and the Liberal Democrat vote 2.0 standard deviations below their respective poll predictions. Clearly, the minor parties did better than the polls suggested.

13.5. Factors Affecting Polls and Voting

Apart from the problem of selecting a good representation of the population for polling, there are other factors than can distort the poll results. For various reasons — perhaps out of sheer mischief — people may say that they are voting for a party other than the one they intend to vote for. Another factor is that if the polls are indicating a strong lead for a party, then its supporters may not vote on the grounds that their votes will not make any difference. Many countries forbid the taking of sample polls in the week or so before the election is due to take place to prevent the polls themselves from influencing the voting pattern of the electorate.

By and large, polls give a good indication of how people intend to vote, although it is not clear that they serve any important democratic function. Some individuals may, indeed, be influenced by the polls in the way they vote. If a small party with extreme views is indicated as having significant support, then those generally sympathetic to those views may be encouraged to vote for that party, since they would not then feel they were part of a freakish minority in doing so. However, whether or not we would agree with their choice, we must accept that that is what democracy is all about!

There are other influences, to some extent unpredictable, which can affect the results of a general election. In the 1992 United Kingdom general election, the Labour Party, headed by Neil Kinnock, was ahead in the opinion polls and seemed to be heading for a comfortable victory. Just before the election, the Labour Party held a rally in the Sheffield Arena that was conducted in a triumphal, rather noisy, American-style that was jarring to the UK public. In the opinion of many commentators this lost the Labour Party some of its support. In addition, the influential and widely-read tabloid newspaper, *The Sun*, brought out a pre-election edition with the front page showing a large picture of Kinnock with a headline requesting that, if Kinnock won the election that day, then the last person to leave Britain should turn out the lights. *The Sun* is also known for its scantily-clad young ladies who appear on page 3. In that particular issue, the lady on page 3 was not young and very overweight with the caption: "Here's how page 3 will look under Kinnock".

It was thought that this intervention by *The Sun*, intellectually irrelevant as it was, influenced enough votes to give a Conservative victory. In the event, John Major became prime minister, with his party achieving the highest proportion of the popular vote ever obtained by a British political party, albeit with a reduced and rather small majority in the House of Commons. *The Sun* certainly thought that its influence had been decisive. On the day following the election, its headline, in typical *Sun* style, was: "It's the *Sun* wot won it". Subsequently, *The Sun* switched its support to the Labour Party (relabelled "New Labour"), which won the three subsequent elections. Such are the factors that influence democracy!

Problems 13

13.1 A large-scale poll of 3,000 electors shows 1,560 are supporting party A and the remainder supporting party B. What is the probability that party A will get more than 50% of the popular vote?

13.2 In a random sample of 200 baby snakes, 92 were found to be male. What is the probability that for the whole population of snakes more than 50% are male?

Chapter 14

Taking Samples: How Many Fish in the Pond?

. . . few are chosen (Matthew 22:14)

14.1. Why Do We Sample?

Consider that we have some large number of entities — they could be human beings, animals, plants or material objects — and we wish to determine some characteristic that they possess. Some examples of this are:

(i) the average height of 14-year-old boys within a country.
(ii) the distribution of hat sizes of males in the United Kingdom.
(iii) the average household income of those living in social housing.
(iv) the average weight of codfish living in the North Sea.
(v) the number of people watching a particular television programme.
(vi) the proportion of faulty components made in a particular factory.

What is clearly impracticable, if not impossible, is to examine every individual entity to determine the required characteristic. It would be possible, although extremely uneconomic, to measure every 14-year-old boy, or to check every component for its perfection, but it is certainly not possible to catch every codfish in the North Sea.

The technique that is used to estimate the required characteristic of a complete set of the entities is to take a sample, that is, after all, what was described in the previous chapter to estimate the voting intentions of the entire electorate.

14.2.   Finding Out From Samples

In sampling theory, the complete set of entities, whatever it is, be it humans, fish or brake pads, is referred to as the population. This population has some property of which we wish to know the average value $\bar{a}_p$, and the standard deviation σ_p. The technique that we use to estimate these quantities is to measure them for a sample of size n, that is to say, that we take a random selection of n entities for our measurements. The individual values of the quantity a in the sample are:

$$a_1, a_2, a_3, \ldots\ldots\ldots\ldots\ldots\ldots\ldots\ldots\ldots, a_n$$

and from these we can obtain the *sample average*, $\bar{a}_s$, and the *sample variance*, V_s, found by the method described in Section 11.2. Without any other information available, the best estimate that can be made for the average of the whole population, $\bar{a}_p$, is the average for the sample, so that:

$$\langle \bar{a}_p \rangle = \bar{a}_s. \tag{14.1}$$

Notice the notation here: the brackets $\langle\rangle$ mean "the estimate for" the quantity they enclose. Now it might be thought that the best estimate of the variance of the population, V_p, is the sample variance, V_s, but a detailed mathematical analysis shows that it is actually given by:

$$\langle V_p \rangle = \langle \sigma_p^2 \rangle = \frac{n}{n-1} V_s. \tag{14.2}$$

The factor $\frac{n}{n-1}$ is called the *Bessel correction* and is especially important in its effect for small samples. For a sample size of 1,000 the Bessel correction is 1.001 but for $n = 9$ it is 1.125. If the sample size were 1, then the estimate of the population variance would be infinity, implying that it cannot be estimated at all, which is sensible, since there cannot be any variation in a single quantity.

Another matter of interest when taking samples is to know how much uncertainty there is in estimating the mean of the population from that of the sample. Instinctively, we know that the larger the sample is, the more reliable the estimate of the population mean is, but we would like to be able to quantify that reliability. To see how to do this, we first imagine the following scenario: there is a large population of millions of entities and we repeatedly take different random samples of n of them and find the sample means, $\bar{a}_s$, each time. The values of $\bar{a}_s$ themselves would have a distribution, and statistical theory shows that, at least for larger values of n, it would be a normal distribution, or nearly so, centred on the true average for the whole population, $\bar{a}_p$. Now, if the variance of the distribution of the values of $\bar{a}_s$ is small, then it is likely that the estimate of the population mean taken from the single sample would be a good one. Conversely, if the variance of the sample means is large, then one could have less confidence that the mean of the single sample was a good estimate of the mean of the whole population.

Theory shows that if the variance of the whole population is V_p, then the variance of the means of samples of size n is:

$$V_{\bar{a}_s} = \frac{V_p}{n}. \tag{14.3}$$

However, the reality is that when you take a sample that is *all* the information you have, and it is from this single sample that you must estimate the variance of the sample mean. The true value of the variance of the whole population, V_p, is not available. Taking estimated values on both sides of Eq. (14.3), and using Eq. (14.2) as the best estimate of V_p, we find:

$$\langle V_{\bar{a}_s} \rangle = \frac{\langle V_p \rangle}{n} = \frac{1}{n} \times \frac{n}{n-1} V_s = \frac{V_s}{n-1}. \tag{14.4}$$

Apart from being found by analysis, this is a sensible result with the right sort of characteristics. If V_s is small, this is an indication that there is not much variation in the values of a, so that the sample mean is likely to be close to the population mean. Again, whatever the variance of the individual values in the sample, the larger the sample, the more likely it is to give an average close to the population average: the effect of $n - 1$ in the divisor of Eq. (14.4).

14.3.   An Illustrative Example

We will better understand the subject matter of Section 14.2 by applying the results to a numerical example. In selling apples an important factor in setting the price is the size, or weight, of the individual apples. While trees do not provide apples of uniform size, they can be separated by size and sold accordingly. A quantity quoted by a grower would be the minimum average weight of the batch of apples they are providing, and a wholesaler would be interested in checking the grower's claim. This is the basis of the following example.

A wholesaler, buying apples from a grower, selects 20 of the apples at random. The individual weights of the apples, in kilograms, are as follows:

0.152 0.203 0.146 0.137 0.123 0.198 0.176 0.139 0.211 0.155

0.139 0.252 0.162 0.180 0.174 0.224 0.156 0.192 0.150 0.167

(i) What is the mean weight and standard deviation of the weights for this sample?

Adding the weights and dividing by 20 gives:

$$\bar{w} = \frac{3.436}{20} = 0.1718 \, \text{kg}.$$

Adding the squares of the weights and dividing by 20 gives:

$$\overline{w^2} = \frac{0.611084}{20} = 0.0305542 \, \text{kg}^2.$$

Hence the variance of the weights for the sample is:

$$V_s = \overline{w^2} - \overline{w}^2 = 0.0305542 - 0.1718^2 = 0.00103896 \, kg^2$$

and the standard deviation for the sample is:

$$\sigma_s = \sqrt{V_s} = 0.03223 \, kg.$$

(ii) Find an estimate for the standard deviation of the sample mean. The estimate of the variance of the sample mean is given by Eq. (14.4), so that:

$$\langle V_{\bar{a}_s} \rangle = \frac{V_s}{n-1} = \frac{0.00103896}{19} = 0.000054682 \, kg^2.$$

This gives the estimate of the standard deviation of the sample mean as:

$$\langle \sigma_{\bar{a}_s} \rangle = \sqrt{\langle V_{\bar{a}_s} \rangle} = 0.007395 \, kg.$$

(iii) The grower claims that the average weight of the apples they supplied is more than 0.2 kg. Can this claim be rejected on the basis of the sample results?

If the grower's claim were correct, then taking many samples of 20 from their apples would give an average sample mean of 0.2 kg *or more* with a standard deviation estimated as 0.007395 kg. The one sample we have taken has a mean weight of 0.1718 kg, which is *at least* $0.2 - 0.1718 = 0.0282$ kg from the claim. If the claim is true, then the probability that the mean weight of the sample is $0.0282/0.007395 = 3.8$ standard deviations or more from the mean is the area in the tail of a normal distribution, and is found from Table 10.1 to be 0.00007 — so very unlikely. The grower's claim is almost certainly untrue.

14.4. General Comments on Sampling

The kind of analysis that was used to test the grower's claims about their apples could equally be applied to claims about the lifetime of electric light bulbs, or to other situations. In examining the claim about apples, we assumed that if we took many sample means they would form a normal distribution. This would only be strictly true if we took very large, theoretically infinite, samples. Otherwise, for smaller samples, instead of using a table for the normal distribution, we should need to use a table for what is known as the *Student t-distribution*,[a] a distribution that depends on the sample size. For a sample of size 20 the error is not negligible, but neither is it large, so using the normal distribution gives simplicity at the expense of some accuracy. However, for smaller samples the error would be significant.

14.5. Quality Control

It is extremely annoying to a customer if, having bought some product, they find it to be faulty and have to return it to have it replaced. It also represents a cost to the manufacturer of the product, who has made an article that yielded no income. For the manufacturer there is a fine balance to be struck. At one extreme, they could check every article before it leaves the factory; this would ensure that there were no returns, but the expense of doing this would have to be passed on to the customer, make their product less competitive and so reduce their sales. At the other extreme, they could check nothing, but then if the manufacturing process developed a flaw through, say, a machine tool becoming defective, then they might not detect this for some time. Then they would suffer the double

---

[a]"Student" was a pseudonym used by the statistician W.S. Gossett who derived the t-distribution in 1908.

loss of needing to replace many articles and also of customer confidence in their product, which would adversely affect future sales. They need to implement something between these two extremes: something called *quality control*.

To illustrate the basic concept of quality control, we consider a factory turning out 1,000 bicycles a day. Experience in the industry suggests that it is acceptable if less than 1% of the bicycles have a flaw, so the manufacturer must monitor their production to keep faulty products to below that level. They decide on a strategy of testing every tenth bicycle and using as a sample the last 1,000 tested bicycles for their statistical analysis. They also want to be 99% sure that the 1% fault level is not exceeded. The question that we now ask is: "What is the maximum number of failures in the current sample of 1,000 bicycles, such that there is a probability of 0.01 or less that 1% or more of the bicycles are faulty?" It is worth reading that last sentence again so that you properly understand the nature of the exercise.

We will deal with this problem in a similar way to that in Section 13.2, where voting intentions were considered. If the measured number of failures in the sample of 1,000 is x, then the estimated proportion of failures is:

$$f = \frac{x}{1,000}.$$ (14.5)

Following the pattern described in Section 13.2, there will be a probability distribution of failure rate centred on f with a standard deviation:

$$\sigma_f = \sqrt{\frac{f(1-f)}{1,000}}.$$ (14.6)

Assuming a normal distribution, the probability of an actual failure rate greater than 0.01 will be the probability of being more than $0.01 - f$ from the mean, or a number of standard deviations:

$$t = \frac{0.01 - f}{\sigma_f} = \frac{0.01 - x/1{,}000}{\sqrt{\frac{x/1{,}000(1-x/1{,}000)}{1{,}000}}}. \tag{14.7}$$

The condition that there should be a probability less than 0.01 of a 1% failure rate is equivalent to having an area in the tail of the normal distribution beyond t standard deviations less than 0.01. From Table 10.1, this is found to require:

$$t > 2.33. \tag{14.8}$$

Finding a value of x that inserted in Eq. (14.7) will give condition (14.8) is best achieved by trial and error: just trying values of x until you find the maximum value satisfying the condition. The values of t found for $x = 3, 4$ and 5 are shown below:

$$x = 3, \quad t = 4.047; \quad x = 4, \quad t = 3.006; \quad x = 5, \quad t = 2.242.$$

The highest value of x satisfying the condition is 4. If the manufacturer finds more than 4 faults in the last 1,000 bicycles sampled, then they should check and, if necessary, overhaul either their manufacturing equipment or their assembly process.

There is some arbitrariness in the quality control system described here. It was decided to use the most recent 1,000 checks on which to base the statistics, and to be 99% sure of not exceeding the target fault rate. Fewer checks would give less reliable statistics, and more would give a greater interval between the time of realizing that something was wrong and the time from the beginning of the sample when things started going wrong. Similarly, if it was one in every 20 bicycles that were checked, this would also increase the time between the occurrence and detection of a manufacturing problem. Getting a good balance is part of the skill of quality control.

What has been described here is much simpler than any system that would be used by a large-scale manufacturer, but it illustrates the general principles involved.

14.6. How Many Fish in the Pond?

Sampling theory can be used in various ways, and here we describe an interesting application. The owner of a large fish pond wishes to know how many fish it contains — at least to an accuracy of about ±50%. They suspect that the number is of order 3,000, but it could be half as much or even twice as much as that. It is not possible to count the fish individually, so what should they do?

The average lifetime of the type of fish in the pond is about three years, so a process for finding the number that takes about a month will not be bedevilled by excessive changes of the total number by births and deaths. On a daily basis they go to different parts of the pond, net a few fish, and tag them with a light plastic tag attached to a fin that causes the fish no distress. The fish are then released back into the pond. Once they have tagged about 400 fish, they then start a process of again going to different parts of the pond and netting fish, but now they just count the number of fish caught and the number of those that are tagged. Let us see how they find the number of fish in the pond from this procedure.

We take a numerical example to illustrate the process. When they carry out their final netting, they find that of the 300 fish they have caught, 60 are tagged. On that basis, the best estimate of the proportion of tagged fish in the pond is:

$$p = \frac{60}{300} = 0.2.$$

However, from Eq. (13.2), the standard deviation of that estimate is:

$$\sigma_p = \sqrt{\frac{p(1-p)}{300}} = 0.0231.$$

From Table 10.1, we find that, for a normal distribution that we assume here, the probability of being within $2\sigma_p$ of the mean is about 0.95. In this case, the limits of p up to two standard deviations from the mean are:

$$p_{low} = 0.2 - 2 \times 0.0231 = 0.1538 \quad \text{and}$$

$$p_{high} = 0.2 + 2 \times 0.0231 = 0.2462.$$

If the number of fish in the pond is N, and 400 fish were tagged, then the actual value of p is $400/N$ because we know that 400 fish were tagged. Hence the limits found, with a probability 0.95, for the number of fish in the pond are:

$$\frac{400}{N_1} = 0.1538 \quad \text{or} \quad N_1 = 2,601$$

and

$$\frac{400}{N_2} = 0.2462 \quad \text{or} \quad N_2 = 1,625.$$

The most probable number, with $p = 0.2$ is $N = 2,000$.

Problems 14

14.1 A sample of 20 Watutsi adult males gave the following heights, in metres:

1.92 1.97 2.03 1.87 2.10 1.85 1.93 1.89 1.92 2.14

2.08 1.97 1.87 1.73 2.06 1.99 2.04 2.02 1.88 1.97

Find:

(i) the mean height of the sample.
(ii) the standard deviation of the sample.
(iii) an estimate of the standard deviation for the whole population.
(iv) an estimate of the standard deviation of the sample mean.
(v) the probability that the mean height of Watutsi adult males is greater than 2.00 m.

14.2 The owners of a fish pond catch and tag 100 fish. They then catch 100 fish and find that 20 of them are tagged. What is the most likely number of fish in the pond?

Chapter 15

Differences: Rats and IQs

Le peuple n'a guère d'esprit ... (The people have little intelligence ...)
(Jean de la Bruyére, 1688, *Les Caractères*)

15.1. The Significance of Differences

It is believed that there are as many brown rats as there are people in
the United Kingdom, mostly living within city environments. Once
ensconced within an urban environment they are not great travellers;
moving no more than 100 m from their nests if the food supply is
adequate. It is then quite feasible that the gene pools of rats in well-
separated places may differ significantly.

We consider a hypothetical example of comparing the weights
of male rats from London and Manchester. A sample of 400 rats
from London had a mean weight of 595 g and a standard deviation
of 41 g, while a sample of the same size from Manchester had a mean
weight of 587 g and a standard deviation of 38 g. The question we
consider is the likelihood of having that difference, or more, in the
mean weights if the samples were taken from the same population. If
that probability is very low, then we could say that there is evidence
that the populations in London and Manchester are significantly
different, otherwise there would be no such evidence.

As is usual in problems of this kind we make a *null hypothesis* (Sec. 6.2), in this case, that London and Manchester rats are all
part of the same population, so that there is a single population
from which we have drawn the two samples. Theory shows that the

expected difference of the sample means is zero, i.e., if we compare the sample means from a very large number of samples taken from the same population, then the difference from one sample mean to another is sometimes positive and sometimes negative, but averages to zero. Another theoretical result is that the variance of the difference will be:

$$V_{diff} = \frac{\sigma_1^2 + \sigma_2^2}{n_1 + n_2 - 2}, \tag{15.1}$$

where σ_1 and σ_2 are the sample standard deviations and n_1 and n_2 are the sample sizes. For the present example, this is:

$$V_{diff} = \frac{41^2 + 38^2}{400 + 400 - 2} g^2 = 3.916 \, g^2,$$

giving the standard deviation for the difference of sample means, $\sigma_{diff} = 1.98 \, g$. To spell out what this signifies, if we take samples of 400 from the same population, then, from Table 10.1, 68% of the magnitudes of differences of sample means will be less than 1.98 g and 95% of them will be less than 3.96 g.

Really, for this kind of problem, the Student t-distribution, as mentioned in Section 14.4, is appropriate, but for the large samples we are using here, assuming that the normal distribution applies will give little error. The actual difference in the sample means from the two cities is 8 g, or just over 4.0 standard deviations. In considering the significance of the difference, we must consider the possibility of having a difference of four standard deviations *in either direction*, that is, we must find the area in the two tails of the distribution more than four standard deviations from their mean. From Table 10.1, the area in one tail more than 4σ from the mean is 0.00003, so in the two tails it is 0.00006. This is a very small probability, so that the difference in the means of the two samples is unlikely to have occurred by chance and we may conclude that the rat populations in the two cities are *significantly different*. Whether or not one can say that they are *genetically different* will depend on a judgement about environmental factors.

If there is some prevailing condition in Manchester that leads to a smaller food supply, then the difference in mean weights may be due to that factor alone and not to genetic variation.

15.2.   Significantly Different — So What!

The probability curves for the rats' weights, as indicated by the samples, are shown in Fig. 15.1.

The fact that the distributions are different, and that London rats have a somewhat higher mean weight, is clear from this form of presentation. However, what is also clear is that the weights are very similar and the statistical term "significant difference" must not be confused with a judgemental term "important difference". Faced with an average London rat and an average Manchester rat, you would be hard pressed to detect any difference.

We have already indicated that a statistical difference may be the result of environmental factors, for example, a better food supply for rats in London than in Manchester, or it may be due to genetic factors. These factors are usually popularly referred to as,

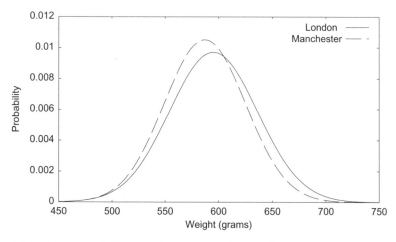

Fig. 15.1.   The probability distributions for the weights of rats in London and Manchester.

respectively, *nurture* and *nature* and, in general, they occur together. If nurture were the sole cause of the difference in weight between Manchester and London rats, then a group of Manchester rats transported to London would have progeny with the same mean weight as London rats.

The human population of the world is subdivided in various ways: by nationalities, by ethnicity, by religion, etc. Some of these divisions are related to genetic differences that manifest themselves through visible characteristics, such as skin colour, hair colour, height and physique. Athletic prowess may well depend on genetic factors, although cultural and environmental factors will also play a role. Thus, people of West African origin seem to do well in track sprint events, while endurance events on the track tend to be the speciality of those from East and North Africa. An Ethiopian sprint champion would be regarded as an oddity, but no one is surprised when an Ethiopian or a Kenyan wins an international marathon race.

While ascribing differences of athletic ability to genetic factors is largely non-controversial, as soon as there is any discussion of differences of ability related to the workings of the brain there is much controversy. An anthropologist who specialized in the study of tribes in New Guinea recounted a story about accompanying some members of a tribe on a hunting trip. On reaching a clearing in the jungle, the headman stated that, since they had last been along that path a few days previously, a large animal had passed through. When asked how he knew this, he replied that some of the vegetation had been disturbed — but the anthropologist looking round the clearing could see nothing of any significance. He decided that the headman was trying to impress him and that really there was no evidence of the passage of a large animal. Such a deception would be safe enough; how could the anthropologist with his Western city background know one way or the other? However, the anthropologist decided to conduct a test. On returning to the village, he spilled a box of matches into a heap on a table and asked the headman to examine it and commit the arrangement to memory. Using a Polaroid camera, he then photographed the pile from several different directions to preserve an accurate record of how it had been. Then, after three days, he gave the headman the same box of matches and asked him to arrange them as he had seen them previously. The pattern of matches set up was not perfect, but it was topologically correct in that the system of matches being under and over others was exactly as in the original pattern. The anthropologist concluded his description of this remarkable feat by noting that: "By any intelligence test we have devised these people would appear to be subnormal; by any intelligence test that *they* would devise *we* would appear to be subnormal."

When a baby is born its brain has little ability to recognize the world in which it lives. It has a few basic instincts — to cry when hungry or uncomfortable, and to suck when hungry and the opportunity presents itself — but otherwise, its brain is almost like a blank sheet of paper on which anything may be written. As the child grows, so contacts are established by synaptic connections in the brain that condition the child for involvement in the society in which it will live and enable it to develop the skills necessary to flourish in that environment. Most of this process of brain development takes place in the first three years of a child's life. A child that gets plenty of stimulation in its early years is likely to be intellectually well-developed in later life; one that is deprived of such stimulation will be handicapped thereafter. The intellectual capacity of an individual, and the main skills developed, will depend to a great extent on nurture — the society in which he or she lives and his or her early experiences — but also on nature, in the sense that genetic factors will also come into play. The New Guinea tribesmen needed and acquired different skills from those of the anthropologist, but there is no evidence that the intellectual capacity of their society was in any sense inferior to that of his.

Within a particular society, homogeneous in a genetic sense, there are often different patterns of child development dependent on social class. However, one can learn little about the genetic quality of individuals from such differences. Parents lacking parenting skills will not enable the full potential of their children to be realized, and these children, in their turn, will often become inadequate parents. This cumulative neglect, sad though it is, does not affect the potential bound up in the genes of the affected individuals and, given the right circumstances, that potential could be realized. Trying to create those circumstances is a difficult, perhaps impossible, task that many governments have tried to tackle.

Testing different social or racial groups may reveal significant differences, in a statistical sense, in their ability to deal with intelligence tests or their ability to track animals, but will tell you nothing of interest about the genetic characteristics of the groups being tested. Such differences as are found are often *statistically* significant but may be of no importance. The child of inadequate parents, who would usually perform badly on intelligence tests, might, with the right upbringing, have become a theoretical physicist. The child, influenced by the teaching of a parent who was a rabid racist, could, with the right upbringing, have become a tolerant high dignitary of the Church of England. Statistics is a useful tool for understanding society — but it is important that we should first understand statistics, otherwise we may draw the wrong conclusions from its application.

Problem 15

15.1 A farmer decides to test the effect of a new feeding regime on her chickens. For the chickens fed on the original regime, the mean weight of 100 eggs is 57.2 g with a standard deviation of 2.1 g. For chickens fed on the new regime, the mean weight of 100 eggs is 57.7 g with a standard deviation of 2.3 g. What is the probability that there could be that difference or more in the means of the two samples if the new feeding regime was ineffective?

Chapter 16

Crime is Increasing and Decreasing

O liberté! O liberté! Que de crimes on commet en ton nom! (O liberty! O liberty! What crimes are committed in your name!) (Madame Roland, quoted by Alphonse de Lamartine, 1847, *Histoire de Girondins*)

16.1. Crime and the Reporting of Crime

Committing a crime involves breaking the law and, since what is unlawful is defined in great detail, it would seem that the presentation of crime statistics would be a completely objective process. Actually this is not so. For one thing, the police exercise a certain amount of discretion in applying the law; for example, if exceeding the legal speed limit on roads by a modest amount was always detected and prosecuted, then there would be millions of criminal acts and prosecutions per day. Mostly such "crimes" are not detected, but even when they are, the police will either not take action because the violation is not at a gross level or, perhaps, they will just issue a warning. When a young offender is caught committing a minor crime, a formal warning may be enough to steer him or her away from further crime, whereas prosecution and conviction may do no more than create another citizen with a criminal record. Another uncertainty in crime statistics is that not all crimes are reported. A minor act of pilfering, say, involving the loss of money or goods worth just a few pounds, might not be reported. The effort of reporting the crime may be more troublesome than the loss itself, coupled with the knowledge that the police are unlikely to devote much effort to solving such a trivial crime when they have more serious matters under consideration. Even serious crime is sometimes unreported. Acts of domestic violence may be unreported, because the victim is too ashamed to reveal the sad state of, usually her, family life. Another serious crime, rape, is notoriously under-reported, because the victim fears that she, or he, will be accused of having been a consensual partner in the act. Proving rape beyond reasonable doubt is often very difficult.

As a result of various government and police initiatives, there are changes over time in both the reporting and recording of crime, and this is one of the factors that confuses any consideration of trends in crime statistics. Rape seems to be more reported than it once was, as procedures have changed and victims are less fearful that they will be exposed to humiliation, both in the initial police investigation and in the courts. Another factor is that criminals may, from time to time, change the direction of their criminality. In recent years, the security of motor vehicles has been improved so that car crime has become more difficult to carry out. A criminal, seeking to fund a drug habit, may therefore be deflected away from car crime towards some other criminal activity.

Crime is a rich source of point scoring for politicians, and an observer of a political debate on crime, in which statistics are bandied to-and-fro, might well conclude that crime is simultaneously increasing and decreasing. Here, we shall look at the official crime statistics in England and Wales over the past few years to see what really has been happening. Based on the facts, should we be more or less fearful about crime, or is the incidence of crime fairly stable? If we take the politics out of crime, we may see it as it really is.

16.2. The Trend for Overall Crime in England and Wales

The British Crime Survey is issued by the Home Office and reports on various types of crime in England and Wales in the recent past.

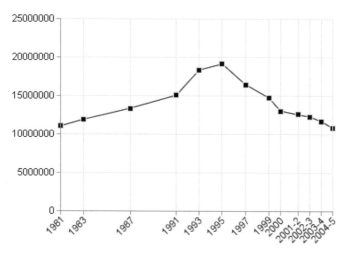

Fig. 16.1.   The numbers of reported crimes of all types in England and Wales from 1981 to 2005 (UK Home Office).

All major types of crime are included in the survey, except fraud and forgery. The figures for overall crime, reported by the survey, are given in the form of a graph in Fig. 16.1.

It will be seen that in the period under report, crime rose steadily from 1981 to 1995, increasing by approximately 80% in that period. After 1995, the total number of offences fell steadily, and in the year 2004–2005 it fell below the 1981 figure.

The rise in crime took place during the Conservative adminis- tration, headed for most of the time by Margaret Thatcher, a point made much of by the opposition political parties. On the other hand, there was a marked decline between 1995 and 1997 (the last years before the Labour election victory of 1997), so the Conservatives can justly claim to have launched the improvement which lasted for over a decade. Labour would claim that since it took power, overall crime has fallen and might neglect to mention that it inherited an improving situation. Such is politics!

The graph shows the facts of the situation but it is not the whole story. Some kinds of crime are of more concern to society than others. Vehicle crime is not trivial — especially to those affected by it — but everyone in society would regard murder as much more serious. If it turned out that burglary and car theft had greatly reduced, but that murder had greatly increased, then an average citizen would feel that everyday life was much more threatening.

16.3.   Vehicle Crime, Burglary and Violent Crime

The incidence of vehicle crime over the period 1981 to 2005, as given by the British Crime Survey, is shown for England and Wales in Fig. 16.2. It mirrors to some extent the pattern shown in Fig. 16.1 for overall crime: a steep rise, more than doubling between 1981 and 1995 and thereafter a fall, starting in the Conservative administration and continuing under the Labour government. A consideration to bear in mind in judging the significance of this graph is a substantial increase in car ownership during the period — by 16% in the decade 1990–2000 alone.

Part of the decrease is probably due to improved security on motor vehicles; many now have immobilizers, which means that they cannot be started by "hot-wiring", i.e., by joining wires behind

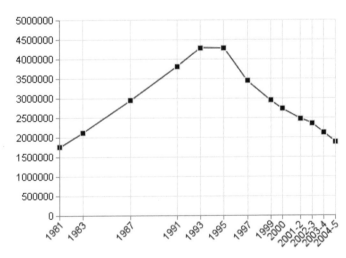

Fig. 16.2.   Vehicle crime in England and Wales from 1981 to 2005 (UK Home Office).

the dashboard, but require the key to get them started. Another factor is that radios, which were once an optional extra, are now standard on all cars, so that there is little market for stolen car radios. Finally, publicity campaigns have persuaded people not to leave valuables — handbags and mobile phones, for example — in view within cars that have been left unattended.

The crime of burglary, stealing by breaking into homes, is less common than vehicle crime, but far more traumatic for the victim. A home is a special place where an individual or a family spend most time, and in which they normally feel secure. Even the police cannot enter a home without a warrant or some other authorization. A burglary is a violation of that security, and some victims say that they never again quite feel the same about their homes once they have been burgled. There are no such feelings about vehicles; a country at war exhorts its citizens to defend their homes — not their motor cars! Figure 16.3 shows the incidence for burglary in England and Wales from 1981 to 2005, a pattern that by and large resembles those for overall crime and vehicle crime, except that the decline began a little earlier, in 1993.

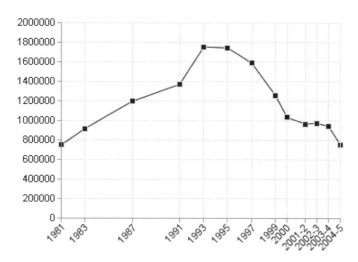

Fig. 16.3.   Burglary in England and Wales from 1981 to 2005 (UK Home Office).

The pattern, as presented so far, is reassuring in that the present trend is downward. However, vehicle crimes and burglary rarely, although sometimes, involve violence and it is violent crime that is of the greatest concern to most people. To have the alloy wheels of your car stolen while you sleep may be annoying, but to be in hospital after being violently assaulted is an experience in a different league altogether.

The category of violent crime includes domestic violence, common assault and mugging, but does not include homicide. The occurrence of violent crime in England and Wales from 1981 to 2005 is shown in Fig. 16.4.

The scale of violent crime, several million per year, is quite surprising, but at least it shows the same decline from 1995 as the other categories of crime so far described. In the year 2004–2005, there were 401,000 cases of domestic violence, 347,000 cases of mugging and the remainder were assaults of one sort or another. A high proportion of the assaults are the regular Saturday night "punch-ups" that occur in town and city centres after too much alcohol has been consumed. However, a citizen who leads a quiet domestic life and

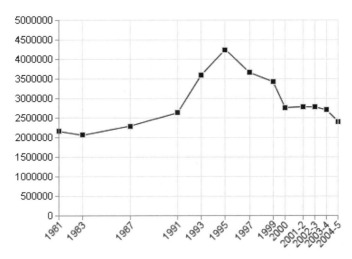

Fig. 16.4.   Violent crime in England and Wales from 1981 to 2005 (UK Home Office).

has friendly neighbours is not very likely to become a victim of violence — although, alas, some do.

16.4. Homicide

Homicide is the category of crime that includes murder and manslaughter, and is the one that most impinges on the public's general awareness of crime. A murder is always reported in the local press and on local radio and television news programmes. A homicide with special characteristics, involving, say, multiple deaths, a celebrity, a child or the use of a firearm will usually become a national news item. This concentration of attention on homicide in the news media leads to the impression that it is a common occurrence in society. The incidence of homicide in England and Wales from 1993 to 2005, as displayed in Fig. 16.5, shows that, on the contrary, it is a very rare crime. The numbers per year are measured in hundreds, whereas the numbers per year of the other types of crime to which we have referred are measured in millions. However, the graph shows a different evolutionary pattern to those of the other types of crime.

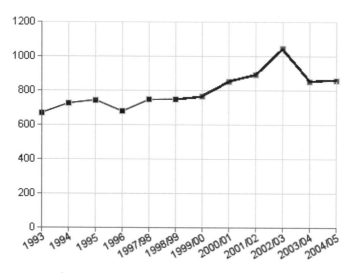

Fig. 16.5. Homicides in England and Wales, 1993 to 2005 (UK Home Office).

The later entries in the graph are a little misleading, because the 2000–2001 figure includes the deaths of 58 Chinese illegal immigrants who suffocated in a lorry bringing them into the United Kingdom, and the 2002–2003 figure is greatly distorted by including all the victims of the mass murderer Dr. Harold Shipman (estimated to be between 150 and 250) in that year, although their deaths occurred over several years. Taking these special factors into account, it is probably true to say that the underlying rate has flattened out at a level less than 10% above what it was in the late 1990s, to something in the region of 850 per year. It is customary when reporting homicide rates in any society to express it as homicides per 100,000 of the population per year. Given the population of England and Wales is about 53 million, this gives a rate of 1.6 per 100,000 per year.

As might be expected, the rates tend to be higher in major cities; the figures for three major cities in the UK are:

London	2.1
Edinburgh	2.4
Belfast	4.4

The Belfast figure is affected by a component of sectarian homicide which is not present to any extent in the remainder of the United Kingdom. However, while any homicide is one too many, it turns out that the United Kingdom is one of the safest countries in the world in terms of homicide. Of all major countries, only Japan has a lower homicide rate (0.9 per 100,000 per year, but accompanied by a huge suicide rate); even Switzerland, generally regarded as a haven of harmony and tranquillity, has more than twice the UK rate. To put the UK city figures in perspective, some quoted figures for major cities in Europe and the United States are:

New York	7.3
Amsterdam	7.7
Lisbon	9.7

Chicago	22.2
Baltimore	38.3
Washington DC	45.8

Despite all the impressions to the contrary, the United Kingdom is a safe place to live — at least as far as homicide is concerned. UK citizens should be concerned about homicide, but it should not overshadow their lives — they are four times more likely to be killed in a road accident than to be the victim of a homicide.

16.5.   Crime and Politicians

There are many issues that affect the lives of citizens of any country: education, law and order, health, employment and taxes, to mention just a few. The facts relating to any of these issues will be complex and interact with other facts in unexpected ways. In this chapter, a numerical account of the incidence of crime in the United Kingdom has been given. But even this picture, objective as it seems, is open to criticism. The British Crime Survey finds out about crimes that have occurred by asking victims of crime (with the obvious exclusion of homicide victims) about their experiences, and their survey includes both crimes reported to the police and those unreported. Their claim is that this form of collecting statistics is more comprehensive than just relying on police figures, and also reveals more about the public's attitudes towards crime. However, some question the validity of this way of collecting crime statistics.

There may be a number of reasons why crime statistics vary from year to year. There will be a statistical fluctuation based on the random pattern of the incidence of crime. If over a long period of time it is assessed that the average rate for a particular crime is N per year, then the standard deviation in that number due to random fluctuations is $\sqrt{N}$ per year. For $N = 2,000,000$, we have $\sqrt{N} = 1,414$ which is 0.071% of the number of crimes committed — a very small fractional fluctuation. On the other hand, if 900 homicides are expected in a year, then $\sqrt{N} = 30$, which is 3.3% of the expected number and a much larger random fluctuation. Because of the numbers involved, with the exception of the homicide figure, the graphs shown in this chapter are little affected by random fluctuations: the trends are real.

Another important reason put forward for changes in the crime figures is changes in the unemployment rate. During the period that crime rates have fallen in the United Kingdom, so has the unemployment rate, and there is an assumption that crime rates will be higher among the unemployed. This was not evident during the slump years of the 1920s and 1930s, but during that earlier period there was less taking of illicit drugs, no television, and so much less exposure to advertising that stresses the attractiveness of acquiring ever more worldly possessions. No individual government can be blamed for a worldwide change in advertising pressures, or for a worldwide depression, so perhaps it is unfair to blame a government for an increase in crime during its tenure of office.

The argument just made as an excuse for the increase in United Kingdom crime from 1981 to 1995 may or may not be sustainable. The point really being made is that a simplistic interpretation of any set of statistics, whether it relates to crime or anything else, may be flawed. We live in a complex world. Politicians, for their own ends, will try to put forward a simple picture that either presents their own party in a favourable light, or the opposition in an unfavourable light. The rule is to try to think for yourself; a useful exercise is to be a devil's advocate and to see if you can make a case for the other side, because then you might see the inconsistencies in the argument being presented to you. One should beware of confidence tricksters, whether they are after your money or your vote.

Problem 16

16.1 The following graph shows the number of road users killed in road accidents in the United Kingdom from 1964 to 2005

(UK Home Office). Estimate from the graph:

(i) the ratio of the number killed in 2005 to the number killed in 1964.

(ii) the fraction of those killed who were pedestrians in 1964, 1982 and 2005.

Suggest some reasons for the general decline in the figures over the period of the graph.

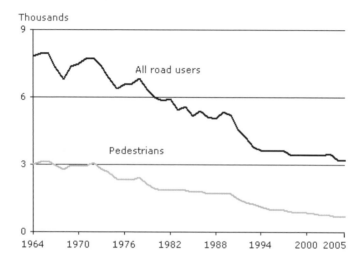

Chapter 17

My Uncle Joe Smoked 60 a Day

...damned tobacco, the ruin and overthrow of body and soul (Robert Burton, 1621, *The Anatomy of Melancholy*)

17.1. Genetics and Disease

At the beginning of summer many plants enter their period of reproduction and release vast quantities of pollen. For most people this is a matter of no consequence, but for an unfortunate minority it presages a time of great misery and discomfort. These are the sufferers from hay fever. They sneeze violently and continuously, their nasal passages become blocked with mucous, their eyes become red and inflamed, and they have difficulty with breathing. Some relief is obtained by taking antihistamine tablets, but they have the unpleasant and potentially dangerous side effect of causing drowsiness. Unfortunately, the onset of the hay fever season coincides with the period when many young people take important examinations and, unless some allowance is made for their condition, hay fever sufferers may have their futures blighted by performing below their true capability.

Hay fever is recognized as being of genetic origin and its occurence in families is often associated with eczema and asthma. It is a prototype of a genetic disease of which there are many other examples. Breast cancer in women is now recognized to have a strong genetic association and, indeed, the genes associated with the condition, labelled as BRCA1, BRCA2 and BRCA3, have now been detected. Researchers have stated that 60% of women with the BRCA3 gene will develop cancer before the age of 70. Nevertheless, only 5% of cancers are associated with the presence of the aforementioned genes, so there are clearly other factors at work.

Cancer is a condition in which some cells of the body run out of control and repeatedly divide, forming a tumour. If the condition remains localized, then the tumour is benign and removing it, by surgery or locally-applied high radiation doses, essentially cures the condition. However, if the tumour cells migrate through the body affecting other tissues, then the condition is malignant and much more difficult to handle. Cancer may be caused by various agencies, but basically all these have the effect of damaging some of the DNA in the cell. This could happen by exposure to radiation — for example, by excessive sunbathing — or by the action of carcinogenic chemicals, of which there are many.

The breast cancer example shows that not all those with a genetic predisposition towards a particular disease will suffer from that disease, and also that those without the predisposition sometimes will. There is an element of chance in determining whether or not a particular individual is affected. A potentially cancer-inducing agency may, or may not, damage cell DNA to cause a cancerous cell. The probability that it will do so would depend both on the degree of exposure to it and on the propensity of the individual to suffer such damage. It can happen that someone who has a genetic tendency towards some form of cancer, but who is little exposed to the agencies that cause it, may never contract the disease. It is even possible that such a person could be heavily exposed to a carcinogenic agency and *not* contract the disease — but they would have to be lucky. By contrast, someone without any genetic disposition towards cancer could well contract the disease under any level of exposure, but with the risk increasing with exposure.

It is against this background that we will discuss the relationship between smoking and smoking-related diseases.

17.2.   The Incidence of Smoking in the UK

The smoking of tobacco originated with the aboriginal inhabitants of North America and was introduced to Europe by various explorers during the sixteenth century. It probably came to England through Sir John Hawkins, or his crew, in about 1564, but was popularized by Sir Walter Raleigh, a rakish adventurer and a favourite of Elizabeth I — but somewhat less favoured by her successor James I, who ordered his execution. During the eighteenth and nineteenth centuries the idea was prevalent that smoking was, in some way, beneficial to health — perhaps because the smell of tobacco was preferable to the prevailing smells of the time.

During the early part of the twentieth century smoking cigarettes was regarded as a rather sophisticated thing to do, and films from the pre-Second World War period show people smoking (and drinking) quite heavily, with languid wealthy young ladies doing so with the aid of long cigarette holders. However, in the latter half of the twentieth century evidence began to emerge of a connection between smoking and various diseases — notably lung cancer, but also many others. At first, the major cigarette companies fiercely disputed this evidence, saying that it was not well founded. Anti-smoking organizations even suggested that the companies were withholding information connecting smoking and lung cancer that their own research had revealed. The companies were, to a great extent, initially supported by smokers themselves, who were in a state of denial. Often they would use irrational arguments to justify continuing to indulge in their harmful drug habit, of the kind: "My Uncle Joe smoked 60 a day for the whole of his life and lived to be 101". Sometimes the same Uncle Joe would be cited by other nephews and nieces as a reason to drink alcoholic beverages to excess because that is what he did.

Eventually, the cigarette companies had to give ground as responsible governments all over the world took steps to encourage people to give up smoking. The advertising of tobacco products was banned on television and in other media. In addition, the tobacco companies had to put warnings on the packets of their products, so one has the bizarre spectacle of individuals giving large sums of money in return for packets bearing the words: "Warning — cigarettes can kill". Smoking was banned in many public places, for example, in theatres and cinemas. Many of the early films, in which smoking was portrayed in such a glamorous way, had to be viewed through a thick haze of tobacco smoke before this ban. Restaurants designated smoking and non-smoking areas and many airlines prohibited smoking, even on long-haul flights. In 2004, the Republic of Ireland banned smoking in what had been, until then, the smoker's paradise: the public house. The various component countries of the United Kingdom followed suit in 2006 and 2007, banning smoking in all public places.

All this pressure against smoking gradually had an effect, and the proportion of smokers of both sexes steadily decreased in the United Kingdom: from 39% of all adults in 1980 to 21% in 2010. Figure 17.1 shows the variation in the proportions of men and women who smoked from 1980 to 2008, based on data published by the UK Office for National Statistics. The United Kingdom government has set a target of halving the 2010 level of smoking by 2020.

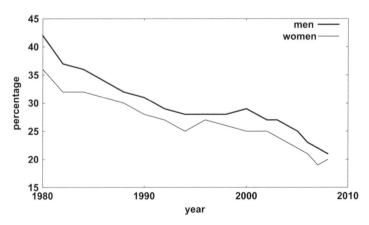

Fig. 17.1.   The incidence of cigarette smoking in the United Kingdom (UK Office for National Statistics, 2009).

By definition, reducing the overall proportion of those smoking will reduce the incidence of smoking-related illness. By and large, the argument against smoking, in terms of the health hazard that it imposes, has been won; those continuing to smoke either accept the risk involved, perhaps because they are so addicted that they cannot stop, or because they are completely unable to understand the arguments that have been accepted by the majority. For them the "Uncle Joe" example is the only one they can understand.

Nevertheless, Uncle Joes do exist, so we will now consider why it is that Uncle Joe got away with his reckless lifestyle, while other clean living, non-smoking individuals die young of lung cancer, or some other disease normally associated with smoking.

17.3. The Smoking Lottery

Unless there is a family history of smoking-related illness or genetic screening of some kind, an individual cannot know what his or her sensitivity is to tobacco smoke or, indeed, to any other carcinogen. There is no way of avoiding *some* exposure to potential carcinogenic hazards. Background radiation is everywhere, and is particularly high in areas where granite exists in the rocks, such as in parts of Cornwall, or where granite is heavily used as a building material, as in Aberdeen. While smoking is still a legal activity, there is the hazard of secondary exposure when in the presence of those smoking. This is especially true in the home if some, but not all, of the family smoke, and young children are particularly put at risk by the smoking of one or both parents. Even with the best of filters, the exhaust systems of cars emit some carcinogens and occasionally one sees badly maintained vehicles illegally emitting black exhaust fumes that are both unpleasant and highly toxic.

Given that there is no escape in the modern world from some exposure to carcinogenic hazards, and also that, in general, one's sensitivity to such hazards is unknown, the prudent course of behaviour is to avoid any additional exposure, such as smoking.

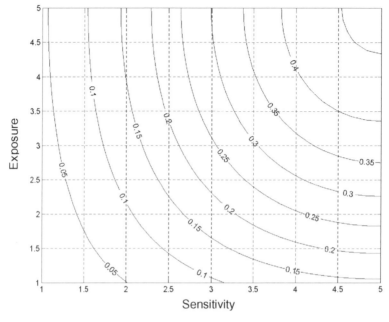

Fig. 17.2. A notional representation of the probability of acquiring a carcinogen-related illness as a function of exposure and sensitivity.

The probability of carcinogen-related illness increases with sensitivity and increases with exposure, although exactly quantifying the risk is not easily done. Just as an illustration of what the situation could be, there is given in Fig. 17.2 a notional representation of the dependence of the probability of acquiring a carcinogen-related illness related to exposure and sensitivity. The sensitivity is given on a scale of 1–5 on the basis that nobody can have so little sensitivity that they are immune from such illnesses, and exposure is also given on a scale of 1–5 on the basis that there is a minimum exposure that cannot be avoided.

From the figure, it will be seen that, for someone with a sensitivity rating of 2.0, the probability of getting a carcinogen-related illness goes from 0.05 at minimum exposure to about 0.16 at maximum exposure: an increase factor of more than 3 in risk. Someone who is maximally sensitive goes from a probability of 0.15 with minimum exposure to about 0.45 with maximum exposure: again an increase factor of 3. Uncle Joe may well have had maximum exposure but if he had minimum sensitivity, then the likelihood that he would acquire a carcinogen-related illness would only be 0.05: about a 1 in 20 chance. However, if there were 20 people like Uncle Joe, both in sensitivity and exposure, then it is very likely that one of them would become ill. Which one would succumb, and which ones would not, could not be determined *ab initio*, even if the facts about sensitivity and exposure were known. That is the smoking lottery.

The story about smoking and illness is one that is now well understood and, essentially, will not change with time, even though the exact nature of the risks involved may be quantifiable with greater accuracy as the results of further research are made available. In the comedy film *Sleeper* (1973), starring Woody Allen, a nerdish shopkeeper is awakened from a cryonic state after 200 years into a world where many beliefs and practices have changed. In one scene, he is being examined by hospital doctors who prescribe a course of smoking cigarettes to improve his health. Smoking to promote health may make good cinema humour but is not, and never will be, a reality.

Problem 17

17.1 From the graph in Fig. 17.2, find the probability of an individual acquiring a carcinogen-related illness with:

(i) sensitivity level 1.5, exposure level 5.0.
(ii) sensitivity level 5.0, exposure level 1.5.
(iii) sensitivity level 3.0, exposure level 3.0.

For someone with a sensitivity level of 3.0, living with a background exposure level of 2.0, the increase in exposure level due to smoking 20 cigarettes a day is 2.0. By what factor has he or she increased his or her risk of acquiring a carcinogen-related disease by deciding to smoke?

Chapter 18
Chance, Luck and Making Decisions

Has he luck? (Napoleon I, 1769–1821, when appointing a new general)

18.1. The Winds of Chance

In Edward Fitzgerald's famous translation of *The Rubáiyát of Omar Khayyám* (1859) there is a verse that runs:

'Tis all a Chequer-board of Nights and Days
Where Destiny with Men for Pieces plays:
Hither and thither moves, and mates, and slays,
And one by one back in the Closet lays.

This verse summarizes the beliefs of many people about the nature of life: a series of chance happenings that toss you this way and that, sometimes for good and sometimes for ill. In primitive societies this idea of being helpless, coupled with ignorance about the way that nature works, led to superstitious beliefs in the supernatural. The idea arose that there were all-powerful agencies, interpreted as gods, who shared many of the emotions and characteristics of humankind — anger and generosity, for example — and would act accordingly. If a volcano erupted and life was lost, then this was a sign that a god was angry and was venting its feelings by killing some of those that had offended it in some way.

In most early societies, the idea was established that there was a multiplicity of gods dealing with different aspects of human affairs, such as war, love, fertility, prosperity, etc. There were two main reactions to this belief in a panoply of gods. The first was to try to modify the way that the gods felt, and would therefore react, by prayer and supplication and by making offerings of one sort or another. In the more advanced societies, temples were built and a priesthood was established, which included experts who could act as intermediaries between the layman and the all-powerful gods. They would arrange and supervise the recital of prayers, the offerings of goods, and the carrying out of sacrifices, usually of animals, but also of human beings in some societies. The second reaction was to try to predict what the gods had in store, so that one could take some kind of remedial action. A famous example of this, from Greek culture, was the Oracle of Apollo at Delphi, a temple in which Apollo spoke through the Pythia, a human priestess, and offered guidance to those who came. The pronouncements about the future were usually a bit vague (a characteristic of many religious texts) and required further interpretation. The Romans seemed to favour an investigation of chicken entrails to predict the future. In Shakespeare's *Julius Caesar*, a soothsayer, a person who can foretell the future, warns Caesar to "Beware the Ides of March", but one is not told how he derived his prediction of danger. The idea that some people can predict the future has given rise to a host of terms for describing such people, for example, in addition to soothsayer, there are the terms *seer, visionary* and *fortune teller*. Belief in powers of prediction has declined greatly in the modern world, but there are still those who use tarot cards, astrology, tea leaves, palmistry or a crystal ball to attempt to find out what the future has in store for them.

One of the great scientific discoveries of the last century, at least in a philosophical sense, referred to as the *Heisenberg uncertainty principle*, is that there is no way of precisely defining the future state of the universe, or any part of it. This is in contrast with the deterministic view that existed at the end of the nineteenth century, when it was believed that if the position and motion of every particle in the universe were known at any instant, then the future development of the universe would be completely defined. The uncertainty in behaviour most strongly manifests itself in the behaviour of particles of very small mass, like electrons. The uncertainty principle indicates that it is impossible to know *exactly* where a particle is and how it is moving, i.e., its velocity; if you try to locate its position *exactly*, then this is at the expense of having any idea at all of how it is moving, and if you try to find its velocity *exactly* then you have no idea at all where it is. The Heisenberg uncertainty principle actually quantifies the relationship between the uncertainty of position and the uncertainty of velocity. Theoretically, the uncertainty principle applies to bodies of any mass, but for large bodies of the type we meet in everyday life it has no practical importance. Snooker players do not have to worry about the uncertainty principle in planning their next shot.

The basis of the uncertainty principle is that in trying to determine the position and velocity of a body, you have to disturb it in some way. Finding the position of a body requires it to emit or reflect radiation for a detector to record. The most precise definition of position is obtained if the radiation has a very high frequency, but then the radiation has a great deal of energy associated with it and, in its interaction with the body, will cause the body to move in an unpredictable way. Basically, the principle says that if you go through the process of finding the characteristics of a body, then you influence that body and change its characteristics. Similarly, in relation to human affairs, any attempts to find out about the future will influence and change that future. If Calphurnia, Caesar's wife, had her way, Caesar would not have gone to the Senate on the Ides of March. If one were to accept Shakespeare's version of Roman history, then Caesar would not have been assassinated and human history would have been changed. It seems that the flow of events that dictates both the life of an individual and the destiny of a society are unpredictable and uncertain. As the *Rubáiyát* says, we are moved "hither and thither". By making sensible choices we can minimize the harm that we experience, but we cannot control our lives completely.

18.2. Choices

Every day, people make decisions of one sort or another. They are mostly trivial and sometimes involve a choice of alternative actions, i.e., to go shopping or not to go shopping, and sometimes a choice of several different possibilities — for example, deciding on which resort to choose for a holiday. In Chapter 6 we examined another kind of decision-making: when a doctor decides on a course of treatment, based on some diagnostic information that leaves the patient's condition not uniquely defined. In that case, we were able to assign numerical values to the probabilities of particular conditions, and to the probabilities of the effectiveness of various drugs, and so identify the course of action *most likely* to lead to a favourable outcome. Nevertheless, in a decision based on the calculation of probabilities there is the possibility, or even certainty, that from time to time the wrong decision will be made. In such circumstances, if it turned out that the treatment prescribed by the doctor was not the best for the patient's actual condition, it would be unfair to criticize the doctor or to say that he or she had been professionally incompetent.

Any decision, no matter how trivial it seems, can have unexpected, and sometimes tragic, consequences. The trip to the supermarket by car may be marred by a traffic accident. How many times have those involved in such incidents uttered the words, "If only I had decided to do it tomorrow!"? People who chose to have a holiday in Phuket, Thailand, over the Christmas period in 2004 could not have anticipated or imagined the tragedy of the Asian tsunami that claimed more than a quarter of a million lives. How many relatives and friends of victims have said, "If only they had decided to go somewhere else!"?

We see from the above that in making decisions there are two types of outcome: those that are predictable and those that are not. The latter category is almost infinite in its range of possibilities. When going to the supermarket, almost anything can happen from meeting an old school friend, spraining an ankle, having a traffic accident or winning a prize for being the store's one-millionth customer. If you let your imagination flow, you can fill several pages with possible outcomes of widely varying likelihood. By contrast, by their very nature, predictable consequences are usually bounded. However, one cannot accuse of foolishness a doctor making a wrong decision based on an estimate of probabilities, or a holidaymaker who travels to the scene of a tsunami. That is what life is like, and it cannot be any other way.

18.3.   I Want a Lucky General

There are those who move majestically through their lives, with everything going in their favour, and others who experience one mishap after another. From a statistical point of view, what we are seeing here are the extremes of a distribution of what we might call *good fortune* or *luck*. It was Napoleon who, when confronted with the curriculum vitae of candidates for a vacant general officer post, pushed them aside and said that what he wanted was a lucky general. Rationally, that statement did not make sense. A winning streak at the roulette table would not have any influence on the outcome of future spins of the wheel and, equally, someone who had been lucky in the past was not more likely than anyone else to be lucky in the future. Nevertheless, where decision-making does not depend on mechanical devices like a roulette wheel, there may be people who appear to be lucky in that they make many more good decisions than bad. Actually, such "luck" may just be a subtle manifestation of more skill. The golfer Arnold Palmer is reported as having said, "It's a funny thing, the more I practice the luckier I get". So it is that a lucky general may just be one who has a better appreciation of the lie of the land and so disposes his forces more effectively.

Sporting fixtures very often give unexpected results, with the "underdog" individual or team winning, and sometimes winning well. That is just one aspect of the variation that occurs in all human affairs. In 1985, in the fourth round of the FA Cup competition, the third division club York City beat one of the giants of English football, the first division side Arsenal, 1 – 0. Not only that, but they beat them by outplaying them. The game was played at York, but, nevertheless, the outcome was almost unbelievable. If York City and Arsenal had played 100 games of football, it is likely that Arsenal would have won 99 times, but on that day it was the 100th game. Were York City lucky that day? In a sense they were; presumably there was a concentration of factors operating in their favour and against their opponents. The Arsenal team was in no sense at fault — it was a day when they were at the low end of their performance distribution and their opponents were at the high end of theirs.

What we learn from all this is that in all human affairs there is an element of chance that may affect either performance in achieving certain objectives or the outcome of decision-making. Members of a mature and thoughtful society will recognize this and accept the ups and downs of life for what they are: the manifestation of entirely unpredictable statistical fluctuations of outcomes in particular situations. If rainfall had been low, so that the levels of reservoirs had substantially fallen, then a prudent government minister would order restrictions on the use of water — to the great inconvenience of those affected. If one month after the restrictions were imposed there was a period of heavy rain that quickly replenished the reservoirs, then it may be possible to show that the restrictions had been unnecessary. In the nature of the shallow thinking that accompanies most politically-motivated judgments, the minister would then undoubtedly be accused of having made an unnecessarily premature decision. With the benefit of hindsight decision-making is much easier, and is extremely easy for those that do not have to make the decisions in the first place. Actually, a government minister who ignored the drought by imposing no restrictions, and whose decision was rescued by the subsequent rain, would be worthy of dismissal, despite the fact that his course of action had less inconvenienced the public. He might be a lucky minister on *that* occasion but it would be unwise to put one's trust on the future luck of such a reckless minister. In the next decision he makes his luck might run out and then a disaster could follow his lack of judgment.

18.4.   To Fight or Not to Fight, That is the Question

The world is a very dangerous place because of the nature of its inhabitants. There seems to be an inbuilt tendency in human beings towards violence based on a tribal instinct. However, unlike the small-scale conflicts that went on between tribes in primitive societies, the conflicts of the modern world encompass a huge range of scales. When the football teams Glasgow Celtic and Glasgow Rangers meet, there is an antagonism between the supporters, based on a religious and political divide that has its origins on the island of Ireland. The skirmishes that sometimes accompany such games are unpleasant, sometimes tragic, but are not a threat to humankind in general. Curiously, when Scotland plays England at football, the erstwhile rival fans of the Glasgow teams are united in their opposition to the English tribe. The tribes are not static but can reorganize themselves in various ways based on a football team, a country, colour of skin, religion, school, university, etc. Any individual may be a member of several tribes and so may be an ally of another individual in one context and an opponent of the same individual in another context.

The most important decision that any country can make is whether or not to go to war. If the country is attacked and wishes to defend itself, then there is no decision that needs to be made — someone else has made it already. There are many reasons for going to war as an action rather than as a reaction. Some local wars are caused by border disputes, often between ex-colonial countries, where the colonial powers had created artificial frontiers that did not respect traditional boundaries. In the past, many wars were based on imperialism: the wish to extend the territorial possessions of a country. That was the motivation of the many wars that established the colonial empires of, for example, the United Kingdom and France. Such empire building is now outmoded, a thing of the past. Another reason for declaring war could be based a question of principle. The United Kingdom and France declared war on Germany on 3rd September 1939, because Germany had invaded Poland and the United Kingdom and France had treaty obligations with Poland to defend its integrity. There may have been other background reasons, but, if so, they were not the ones advanced at the time. The Falklands conflict of 1982 might also be claimed to have been a war based on principle, since the economic benefit to the United Kingdom was far exceeded by the cost of reoccupying the islands and the potential dangers that such a military expedition involved. Finally, there are wars based on attempts to obtain economic benefit; the combination of unstable regions of the world plus the presence of oil in those regions has proved to be a very combustible mixture.

The most significant conflict at the beginning of the twenty-first century was the invasion of Iraq and the overthrow of Saddam Hussein. There are many overtones associated with this conflict. Was the invasion of Iraq legal or illegal? Were there good reasons, based on intelligence sources, to believe that Iraq had, or was developing, chemical or biological weapons? Was the public *deliberately* misled about the perceived nature of the threat posed by Iraq? Of course we *now* know that Iraq did not have available stocks of chemical or biological weapons but the general belief at the time, even in those countries that opposed the war, such as Russia, France and Germany, and the United Nations itself, was that it did. There was good reason for such a belief. Iraq certainly had once possessed chemical weapons and had used them against its own population and against Iranian soldiers in the Iran–Iraq war that raged from 1980 to 1988. In this war, the Iraqis were largely supported by a United States that saw the Iranian Revolutionary government as a threat to its oil supplies; for the Iraqis it was more based on a border dispute somewhat fuelled by traditional antagonism between Persians and Arabs. Iraq never fully accounted to the United Nations for the stock of chemical weapons it was known to have had in 1991, although at some time they must have been destroyed by the Iraqis themselves. The UN weapons inspection teams, headed by Hans Blix, often complained of obstruction by the Iraqi authorities, and one must wonder why they were being obstructive, since there was nothing to hide. Given the character of Saddam Hussein, his history of violent behaviour and his reputation for deviousness, it seems remarkable that Iraq *was* free of chemical weapons in 2003 — but so it was. That we now know, but it was not known for certain in 2003 and there were good reasons to think otherwise.

The United Nation's position on Iraq was basically dominated by two blocks: the first headed by the United States and the United Kingdom, which wanted strong and immediate action, and the second by France, Russia and Germany, which urged that the UN inspection teams should be given more time to hunt for weapons. The general view is that the position of the United States and the United Kingdom was mainly dominated by a concern about oil supplies, coupled with a deep distrust of Iraq after that country's invasion of Kuwait in 1990. On the other hand, France and Russia were the countries with the strongest commercial links with Iraq, which owed both of them a great deal of money. Of the major players in this crisis, Germany alone could be seen to be taking a principled stand without any discernible ulterior motive. Resolution 1441, passed by the United Nations, had concluded in Section 13 that: "... the Council has repeatedly warned Iraq that it will face serious consequences as a result of its continued violation of its obligations" (United Nations Document S/RES/1441, 2002). The United States and United Kingdom clearly believed that the General Assembly would prevaricate endlessly and that the time had come to offer those serious consequences as a real possibility, and so assembled forces on the borders of Iraq.

The rest is history as they say. Iraq was invaded successfully, but the follow-up to the invasion by the coalition forces was pathetically inept and the situation developed into a state of chaos, with an influx of foreign terrorists, armed resistance to the coalition forces and, of even greater significance, conflict between different Iraqi groups, based on antagonistic branches of Islam, leading to a large loss of life. An elected government was formed after a successful general election, but the long-term future of Iraq was for some time uncertain.

Both in the United States and the United Kingdom, the tide of public opinion, which at first had supported the war in Iraq — more so in the United States than in the United Kingdom — later swung decisively in the opposite direction, with strong and vituperative criticism of both the president and the prime minister at the time. There can be no doubt that the outcome of the war was not as had been expected — but, as we have previously stated, hindsight is a wonderful guide to policy. Putting aside all questions of the veracity of the leaders or the quality of their intelligence sources, we now try to construct a theoretical framework to assess whether or not there was any rational justification for the path taken by the United States and the United Kingdom.

18.5.   The Mathematics of War

For many people the rights and wrongs of the Iraq War are simply matters of the legitimacy of the action, taken without specific United Nations sanction, and the fact that the justification given for invading Iraq turned out to be invalid. They would argue that even if the war had produced only benign outcomes, it created a dangerous precedent in which the one superpower then on Earth overrode the primary agency for international order, the United Nations, to pursue its own path. Many countries in the United Nations believed that if action were to be taken, then it should have been taken later with the approval of the whole international community, if that were possible. There is strength in that argument. However, there is a counterargument that, with the knowledge that existed at the time the invasion was carried out, there was a possibility of a catastrophic outcome if nothing at all had been done by the international community, and that the international community was far too divided to decide on a course of action. Here we shall sidestep all these questions of legitimacy and morality — important though they are. We shall consider the question of whether to invade or not to invade, by looking at the probabilities of postulated favourable or unfavourable outcomes. This is a notional exercise to illustrate the principles involved in decision-making, as we discussed in relation to medical treatments in Chapter 6. It is not intended as either justification for invading Iraq or for not doing so.

The decision tree for this exercise is presented diagrammatically in Fig. 18.1, where WMD stands for *weapons of mass destruction*, the United Nations term for what it was seeking in Iraq.

Several things must be said about this decision tree before we contemplate using it. First, where alternative possible outcomes deviate from the single point, then the sum of the probabilities must be 1, since the outcomes are mutually exclusive. Thus:

$$p_1 + p_2 = 1; \quad p_3 + p_4 = 1; \quad p_5 + p_6 = 1; \quad p_7 + p_8 = 1; \quad p_9 + p_{10} = 1.$$

The actual values of the probabilities will be a matter of individual judgement; inserting different probabilities will result in different conclusions. Last, the nature of the good and bad outcomes will also depend on individual assessments of what the reactions

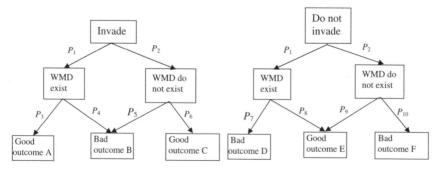

Fig. 18.1.   A decision tree for deciding whether or not to invade Iraq.

of participants — the Iraqi people, foreign terrorists and the Iraqi government — might be. Here we suggest some probabilities and outcomes; readers should insert their own ideas to see what the decision tree suggests. As has been mentioned previously, there are an infinite number of unpredictable outcomes.

Based on general expectations by the United Nations and by individual countries, $p_1 = 0.8$ ($p_2 = 0.2$).

Good outcome A	The WMD are destroyed, a tyrannical government is removed, and a democratic government is established. Iraq becomes peaceful.
Bad outcome B	There is local resistance to the coalition forces, sectarian strife ensues and foreign terrorists exploit the chaotic situation.
Good outcome C	As for A, except that there are no WMD to destroy.
Bad outcome D	Emboldened by the lack of United Nations action, the Iraqi regime use WMD and intermediate-range missiles to threaten or bully its neighbours. Iraq restarts nuclear weapons research and uses the threat of a chemical and biological attack on Israel, delivered by rockets, which would provoke a regional war, to deter intervention by Western powers. A war in the Middle East involving several countries would create world economic chaos when oil supplies were disrupted. This scenario might not appear for a few years and would be difficult to counter.
Good outcome E	The Iraqi regime does not attempt to either begin or expand WMD production and settles into peaceful coexistence with its neighbours. It also stops persecuting opposition groups within Iraq.
Bad outcome F	Despite having no WMD, Iraq continues to threaten its neighbours, confident that the United Nations will not physically intervene.

In view of the violation of human rights and international laws by Saddam Hussein's regime, in considering outcomes A, B and C, a bad rather than a good outcome would be expected. However, there was never nationwide support of the regime from the Iraqi people, so the expectation of a good outcome might be better than expected. This gives the judgment $p_3 = 0.4$ ($p_4 = 0.6$) and $p_5 = 0.6$ ($p_6 = 0.4$).

With the assumed possession of WMD and taking into account Iraq's previous wars with its neighbours, bad outcome D, or some variant of it, seems more likely than the alternative outcome E. We take $p_7 = 0.7$ ($p_8 = 0.3$). Without WMD, Iraq would be less likely to be aggressive, so we take $p_9 = 0.5$ ($p_{10} = 0.5$).

We are now in a position to estimate the total probabilities of good or bad outcomes for both the decision to invade and the decision not to invade.

Invading

Probability of a good outcome is $p_1 p_3 + p_2 p_6 = 0.8 \times 0.4 + 0.2 \times 0.4 = 0.4$.

Probability of a bad outcome is $p_1 p_4 + p_2 p_5 = 0.8 \times 0.6 + 0.2 \times 0.6 = 0.6$.

The sum of these probabilities is 1, as it must be since good and bad outcomes are mutually exclusive.

Not invading

Probability of a good outcome is $p_1 p_8 + p_2 p_9 = 0.8 \times 0.3 + 0.2 \times 0.5 = 0.34$.

Probability of a bad outcome is $p_1 p_7 + p_2 p_{10} = 0.8 \times 0.7 + 0.2 \times 0.5 = 0.66$.

Almost every reader would disagree with some aspect of this analysis or even reject the whole concept of looking at the problem this way. However, what should not be rejected is that it is possible that both options — to invade or not to invade — could more probably give a bad outcome than a good one. Choices in real life are not always between good and bad; often they are between what is bad and what is worse. Whatever one chooses in such a situation, one will be open to criticism — the critics will only see the bad outcome that actually occurred and not the potentially worse outcome of the alternative choice.

What has not been considered in the above analysis is the relative quality of the bad outcomes. Bad outcome D could be so catastrophic that, even if its probability were very low, eliminating it as a possibility might be considered to be of paramount importance. Crossing a quiet country road with your eyes shut might be a fairly safe thing to do, but it is not a sensible thing to do as the potential consequences of a small-probability accident are so bad.

There are many possible scenarios that could be envisaged, but the only one we know the consequences of is the one that actually happened — and that turned out to be very bad indeed. There is a tendency to look for scapegoats when things go wrong. Decisions often have to be made without complete knowledge — knowledge that, with hindsight, the eventual critics will have. Such decisions have to be made on the basis of perceived probabilities of this or that happening, just as one bets on a horse in the light of a probability based on its previous performances. Backed horses do not always win and logically-taken decisions do not always turn out well. That is something an educated citizen should appreciate.

Problem 18

18.1 It could be argued that without the availability of WMD, and with the previous experience of the war in which Kuwait was liberated, Iraq would have followed a peaceful path, so that the probability p_{10} in Fig. 18.3 should be 0. What is the probability of a good outcome in this case if no invasion was carried out? Could this have led to a different decision about whether or not to invade?

Chapter 19

Science and Society

Nam et ipsa scientia potestas est (Knowledge itself is power)
(Francis Bacon, 1597, *Religious Meditations, Of Heresies*)

In Section 3.1 we described the various kinds of probability that are brought to bear in decision-making, which we have called *logical probability, empirical probability* and *judgmental probability*. In the types of decision referred to so far, for example, betting on a horse, it is normally an individual making the decision, although the decision to invade Iraq probably depended on a small group of people at the heart of government. In all these cases the decision-makers had a reasonable understanding of the factors on which their decisions were based.

At the present time there are many important decisions that have to be made by governments worldwide, with the approval of society as a whole, which may affect the future welfare, or even the continued existence, of humankind. The underlying factors on which these decisions depend have a scientific basis and the problem is that comparatively few members of society have a scientific background. For this reason there is a high dependency on the scientific community to rigorously and objectively investigate the relevant factors, to explain their conclusions as clearly as possible to the rest of society, and to give unbiased advice when it is sought. Society must trust the scientists for this process to be effective but, unfortunately, frequently this trust is lacking. Science has conferred many benefits on society in the last few decades — for example, electronic devices of various kinds, new materials and effective drugs and medical procedures — but other scientific contributions, such as atomic weapons, nuclear power and genetically modified crops, have attracted a great deal of hostility and suspicion. It is against this background that we discuss the relationship between science and society.

19.1.   What is Science?

The word *science* is derived from the Latin word *scientia*, meaning *knowledge*. The task of scientists is to attempt to understand the nature of the universe and all it contains, and to determine the laws that govern its behaviour. To this end, scientists carry out experiments and make observations from which they construct theories — for example, about the function of neural networks, the way that stars form or how electrons move within solids. A basic fact about scientific theories is that it is never possible to claim that they are true beyond all possible doubt. The opponents of science, of whom there are many — for example, those who believe in creationism — should not take too much comfort from that statement. Scientific theories of any credibility are evidence-based and are accepted at any time because they agree with all the available evidence at that time. If new evidence is found that shows that the theory is invalid, then this provides the basis for constructing a new theory. This kind of situation is illustrated by the replacement of Newton's laws of motion by Einstein's relativity theory, which showed that Newton's laws are just an approximation — albeit an extremely good one in most normal situations. When sending probes to distant planets, space scientists calculate trajectories using Newtonian mechanics, and they reach their targets with hairline precision. Only for objects moving at close to the speed of light — usually low mass particles which are the fundamental constituents of matter — do we need to take relativity theory into account.

Many conclusions drawn by scientists from their theories are completely non-intuitive and, indeed, may seem to run completely counter to common sense. Relativity theory is a rich source of such conclusions. One of these is that if a man makes a fast round trip to a nearby star then, when he returns to Earth, he will be younger than his twin brother: the so-called *twin paradox*. Fortunately, it is possible to confirm this result using particles rather than human beings. Time has no meaning without change, and the passage of time manifests itself by many types of change, such as the aging of living entities (including us), by the ticking of a clock or by the decay of unstable particles. One such particle is the muon, which at low speeds (ideally at rest) has a half life of 2.2 μs, which means that after every half-life period the number of particles that have not decayed is just one-half the total number at the beginning of the period. Since muons are charged particles, they can be accelerated by electric fields to move at speeds close to that of light, and deflected by magnetic fields to move on a quasi-circular path in a particle accelerator. When they do so, they are found to decay more slowly; for the swiftly moving muons time has slowed down, as seen by the stationary observer. So it is for the travelling twin; as far as the stay-at-home twin is concerned the traveller ages more slowly and arrives home the younger man. This says nothing about the experience of the traveller; he experienced time passing at just the same rate as when he was on Earth, but his clock recorded a shorter journey time than did clocks on Earth.

The point being made here is that the conclusions of science are not always readily acceptable to the non-scientist who is guided by commonsense alone, developed by experience in a world in which objects are large and speeds are slow. There are areas of science of importance to the whole of humankind in which there is conflict between scientists and some non-scientists, or even in which a small part of the scientific community may disagree with the majority. However, before discussing some of the scientifically-based life-and-death issues, we first consider the topic of meteorology.

While weather forecasting is an important activity that illustrates the attitudes of society to scientists, it is not one that critically affects the future of humankind.

19.2.   Meteorology and Butterflies

The state of any region of the atmosphere can be defined by a number of parameters: pressure, temperature, humidity and velocity. If this information were available at closely spaced points covering the whole of the atmosphere, together with the topography of the whole planet, then, taking into account the Coriolis force on the atmosphere due to the Earth's spin, in principle it should be possible to solve a set of equations that would give the state of the atmosphere, and hence local weather conditions, anywhere at any future time.

   The equations that were to be solved were known by the mid-1950s, but the computing power to solve them in a reasonable time was not available; a forecast for the day ahead might take three days to calculate and so was not very useful, although it could be used to verify the validity of the calculation. However, by the end of the 1950s, computer technology had advanced to the point where a one-day forecast took less than a day to compute. The results were based on rather sparse data, and although the programmes usually, but not always, gave a plausible weather prediction, they frequently were quite different from the weather that actually occurred. An American meteorologist working on weather prediction, Edward Lorenz (1917–2008), decided one day to repeat a calculation he had done previously, but to save time he used as his starting point the printed results from the previous run of the programme at some intermediate stage. He was surprised to find that the results he obtained were completely different from those of the first calculation, the difference increasing with simulated time until, eventually, the first and second calculations were not even remotely similar. Lorenz looked into this, and found that the printed numbers used as a starting point for the second calculation had fewer significant figures than the numbers stored in the computer during the first calculation, for example, a number represented as 5.9082573618 in the computer was printed as 5.9082574. This was bad news for meteorologists, since it was clear that tiny changes in the parameters they were using — changes even smaller than the accuracy with which the parameters could be measured — could give a completely different outcome. Short-range forecasts were fairly reliable, but if one tried to forecast too far ahead, then one might as well give a weather forecast based on looking in a crystal ball! From this beginning, there grew a new branch of mathematics called *chaos theory*, which has been found to apply to many areas of science. Typically, it relates to systems where the longer-term behaviour is radically altered by *any* change of one or more initial parameters, no matter how small.

This sensitivity is sometimes metaphorically described in terms of what is called the *butterfly effect*, which states that the disturbance of the atmosphere caused by a butterfly flapping its wings, say on the east coast of the United States, could, in due course, substantially modify weather patterns over the Pacific Ocean, either producing a typhoon or preventing one from forming.

19.2.1. *A demonstration of chaos*

Biologists who study the growth and decline of populations create mathematical models, taking into account food supply, the effect of predation and any other relevant factors. A very simple model is that in a period of one generation the total population changes by a factor r, which can be expressed by the equation:

$$n_{i+1} = rn_i. \tag{19.1}$$

This states that the population of the $(i + 1)^{\text{th}}$ generation, n_{i+1}, is r times that of the ith generation, n_i. It is clear that if the *growth factor* $r < 1$, then the population will steadily decline to extinction, while if $r > 1$, the population will grow without limit. It is impossible in practice for a population to grow indefinitely; large numbers will put pressure on food resources and overcrowding can lead to the propagation of disease. To model these restraining influences, (19.1) has been modified to:

$$n_{i+1} = rn_i \exp(-\alpha n_i),\qquad(19.2)$$

in which the bracketed term on the right-hand side puts a brake on excessive growth.

Calculations with this simple equation, using different values of r, but always taking $\alpha = 10^{-4}$ and initial population size $n_0 = 1{,}000$, give us insights into chaos theory. The first calculation, shown in Fig. 19.1a, is for $r = 2$ and shows steady growth for a few generations, finally reaching a constant population. It is what we would intuitively expect. Initially, the braking term is small, so we get something like exponential growth, but eventually the braking effect just balances the growth effect, so the population stabilizes. Increasing r to 6 (Fig. 19.1b) gives a similar outcome, but with a few wobbles in the curve just before it flattens out.

If r is gradually increased, then at some value this evolution to a constant population breaks down to give a new pattern, shown in Fig. 19.1c for $r = 10$. Now, once the calculation has settled down, the population alternately takes two values, an increase in one generation being balanced by a decrease of equal size in the next. The transition from the long-term behaviour shown in Fig. 19.1b to that shown in Fig. 19.1c happens for infinitesimal changes in the value of r around some critical value.

The next change of pattern, shown for $r = 14$ (Fig. 19.1d), gives an eventual steady state in which there are cyclic variations every four generations. Next, at some value of r, the pattern changes again, as is shown for $r = 16$, giving repeated cycles of six values (Fig. 19.1e). Increasing r further eventually produces an outcome with a cycle of 8 values; cycles contain more and more generations as r is increased, but with the transition values of r getting closer and closer together.

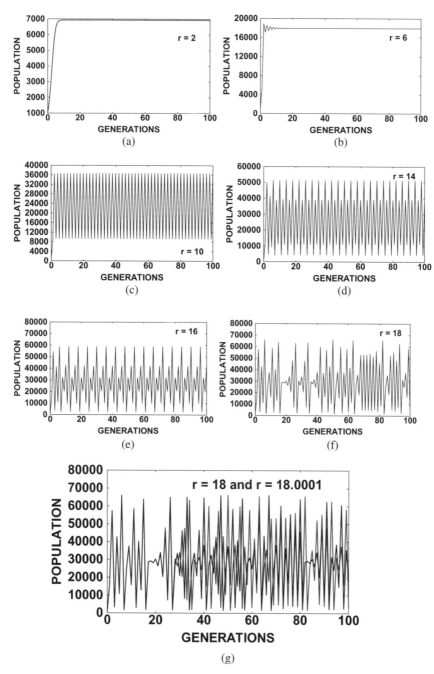

Fig. 19.1. The solutions of Equation (19.2) with $\alpha = 10^{-4}$ and (a) $r = 2$, (b) $r = 6$, (c) $r = 10$, (d) $r = 14$, (e) $r = 16$, (f) $r = 18$ and (g) $r = 18$ and $r = 18.0001$.

The next significant change as r is increased is when the long-term outcome becomes completely chaotic, with no order whatsoever, shown for $r = 18$ in Fig. 19.1f. Now, it was stated that the values of r giving changes of pattern were closer and closer together as r increased. The same is true in the chaotic regime; a small change of r will change the computed curve significantly. Figure 19.1g shows the curves for $r = 18$ and $r = 18.0001$ plotted together. For the first 25 or so generations the curves are indistinguishable, so that the blue curve completely covers the red one. Beyond about 30 generations the two curves are completely different. If the second value of r had been taken as 18.1, then the difference would be evident after about six generations.

If it were possible to estimate the growth factor, r, only approximately, say, as being in the range 17–19, then it would not be possible to make worthwhile predictions of population numbers more than two or three generations ahead.

19.2.2. *Meteorological forecasting*

The weather — how it is and how it will be — has always been of interest, especially in countries like the United Kingdom, where weather can change quickly and unexpectedly. The standard quip about British weather is: "We have four seasons — sometimes all in one day." The interest in weather can be social and non-vital in nature, such as choosing a suitable day for a picnic or knowing what the conditions will be for some sporting event. However, for farmers the weather is a vital factor in determining the success or failure of their crops and the income they derive from them. Knowledge of weather trends is also important to airlines, for whom stormy or foggy weather creates potential hazards, especially in taking-off and landing.

Before the era of fast communications, which we may consider as before the 1830s, when telegraphy using Morse code was established, weather forecasting had to be based on local observations.

Indeed, many people have in their homes a device for doing exactly that: an aneroid barometer, which records pressure and has a moveable pointer from which it can be determined whether the pressure is rising or falling. To a first approximation, high pressure indicates dry weather and low pressure, stormy and wet weather. Finding whether the pressure is increasing or decreasing then gives an idea as to whether the weather is improving or deteriorating. There are other local indicators that are often used. One that is well known, and has a biblical origin, has as its United Kingdom version:

> Red sky at night, shepherd's delight,
> Red sky in the morning, shepherd's warning.

Variants, some referring to sailors rather than shepherds, are quoted in other parts of the world. This folklore depends on the fact that, by and large, weather systems move from west to east. Just before sunset, and just after sunrise, the Sun's rays are seen through the greatest thickness of atmosphere. Passage through the atmosphere preferentially scatters shorter wavelengths of the spectrum, so that sunlight that passes through a great thickness of atmosphere will be reddish in colour. In the morning, if the sky is clear to the east where the Sun rises, the reddish sunlight will reach and illuminate the undersides of any clouds in the west, which eventually will arrive and deliver rain. Conversely, at sunset, if clouds in the east are illuminated, then it indicates that the sky to the west is clear, so that the sunlight could travel without restriction to illuminate the retreating clouds in the east, and this clear sky will give good weather the following day.

Equation (19.2) had just two parameters, and we saw how sensitive the longer-term outcome was to quite small changes in one of them. When it comes to forecasting weather, the number of parameters is extremely large, consisting of measurements of pressure, temperature, humidity and velocity at many locations in the atmosphere. The acquisition of data has been greatly enhanced in the last few decades by commercial aircraft transmitting atmospheric data to meteorological centres worldwide, and by data acquired by satellites

Fig. 19.2. Hurricane Felix (NOAA).

and weather balloons, but even so, the amount of data available is less than what would be considered ideal. Fortunately, there are other sources of meteorological data available: weather satellites. These not only show weather patterns over the Earth's surface, but also how they are moving and so provide useful information to combine with that found by solving the weather forecasting equations. Figure 19.2 is a beautiful example of an image from the GOES-9 weather satellite that shows Hurricane Felix off the coast of North Carolina in August 1995.

The form of Equation (19.2) gives the population at the end of each generation period rather than a continuous variation as time progresses. For meteorological purposes a continuous forecast is required; weather patterns can vary a great deal during the course of a single day. The equations to forecast weather are in the form of *coupled differential equations* in which, at each point of the atmosphere, the rate of change of all the parameters is expressed in terms of the parameters in a region surrounding that point. These equations exhibit the characteristic of chaos in that different runs of the forecasting programme, with changes in parameters smaller than the errors of measurement, give results that depart more and more from each other as the forecasts are extended forward: the characteristic shown in Fig. 19.1g.

Given the chaotic nature of the equations they use, what then can meteorologists do to give useful forecasts? For every parameter they use, say, pressure P, there is an error estimate ΔP, so that the true value of P is probably in the range $P - \Delta P$ to $P + \Delta P$. With modern supercomputers, the time taken to make a forecast, going forward anything from five to sixteen days, is short enough to enable many forecasts to be made within a day. The technique used is to carry out a large number of forecast calculations, in each of which each of the parameters is given a random value within its expected range. The forecasts for a few days ahead will all be similar and have a high probability of being reasonably accurate. Going forward, the forecasts will begin to diverge, but there may be information of a probabilistic kind evident in the results. For example, 20% of the forecasts for the seventh day might indicate rain, and all the temperature forecasts might be in the range 19–21°C. If the forecaster says that it will probably be dry that day, then anyone acting on that information runs the risk of doing the wrong thing, i.e., organizing a picnic on a day when it still might rain. However, this information is better than no information at all, and over a long period one would tend to make more good decisions than bad.

Having obtained this probabilistic information, the question then arises of how to release it to the public. In the United Kingdom, for example, numerical estimates of probabilities are not generally made available. Instead average or expectation values of quantities, such as pressure and temperature, are presented in a graphical form, which makes the information easy to assimilate but not at all detailed. Different agencies use different formats to present forecasts; a typical five-day forecast, as presented to the public, shown in Fig. 19.3, gives a great deal of useful information in a form that is easy to understand. The second column gives a pictorial representation of the general type of weather during the day: on Tuesday the outlook was for dark clouds with occasional sunny spells, but Friday was expected to be a clear and sunny day. The maximum and minimum temperatures might be a guide as to suitable outdoor clothing to

	General summary	Max temp	Min temp	Wind (mph)	Visibility	Pressure (mb)	Relative humidity	Sun index
Monday Sunrise 05:15 Sunset 20:59		31° C	20° C	3	fair	1005	54	moderate
Tuesday Sunrise 05:16 Sunset 20:57		28° C	17° C	5	fair	1009	56	moderate
Wednesday Sunrise 05:18 Sunset 20:55		26° C	17° C	6	fair	1008	51	moderate
Thursday Sunrise 05:19 Sunset 20:54		30° C	20° C	7	good	1011	39	high
Friday Sunrise 05:21 Sunset 20:52		32° C	22° C	12	good	1013	42	high

Fig. 19.3. A typical five-day weather forecast as presented to the public.

wear on the day and the minimum temperature would be of special interest to gardeners, horticulturalists and farmers, especially if it indicated a night-time frost. Variations of low wind speed are usually not of too much interest, but if they reach gale conditions or more severe, i.e., more than about 50 km hour, then structural damage can be caused on land, and ships at sea, particularly small vessels such as trawlers, would need to take precautions or even seek shelter. Wind speed and direction and visibility are also of particular interest to airlines as they affect the safety of the take-off and landing of aircraft. For most of the public the precise values of air pressure and humidity are not of great interest, although high humidity, especially if combined with a high temperature, can lead to a great deal of personal discomfort; in affluent regions of the world where such conditions are common, for example, parts of the United States, air conditioning is usually installed to give comfortable home and work environments. The final column, the Sun index, is very important, but frequently ignored. The ultraviolet component of sunlight is harmful to health, and excessive exposure to the Sun can lead to melanoma and other forms of cancer.

As previously explained, weather forecasts are based on an aggregation of results from a large number of runs of the forecasting programme. If 80% of the runs give dry conditions on a particular day, and the other 20% indicate rain, then the forecaster will indicate the probability of a dry day with some risk of showers (see Problem 1.1). If it turns out to be a very wet day, then the public's trust in the forecaster is eroded, and if there were several unlucky forecasts within a short period, then the public's trust may be lost altogether. In some forecasts in the United States the probabilistic nature of forecasts is specifically indicated by statements such as "chance of precipitation 60%", thus indicating clearly to the public that there are uncertainties in the forecast.

Meteorologists do not help their reputations by sometimes making long-term predictions that cannot possibly have any validity, given the chaotic nature of weather systems. Thus a forecaster in the UK Meteorological Office predicted a "barbecue summer" in 2009 when, in fact, it was rather a wet summer, although with a slightly higher than average temperature. Another occasion when the Meteorological Office's reputation received a severe dent was in October 1987, when southern England and northern France were struck by a massive storm which killed 22 people in England, destroyed a great deal of property and felled a large numbers of trees. The fact that bad weather was on the way was known, but it was predicted to be restricted to a region to the south of the United Kingdom. Up to the evening before the storm struck the predictions being given were that the United Kingdom would not be affected.

19.3. Global Warming

An outstanding example of a science-based issue of immense importance to the future of humankind is *global warming*, the progressive increase in the Earth's temperature, allegedly due to the activity of humankind. It provides a good example of the kind of understanding gap that can exist between scientists and non-scientific members of society. It is a conflict in which probabilities are involved, with the difficulty that we have no way of giving precise numerical values to those probabilities. In Section 18.5, where the pros and cons of going to war were considered, it was possible to establish a tenuous mathematical approach, although there were uncertainties as to what the range of outcomes might be. In dealing with scientific issues there is the similar possibility of phenomena being relevant that we cannot take into account, because we do not know that they exist. Thus, Isaac Newton (1687, *Philosophiae Naturalis Principia Mathematica*) gave as one of his "Rules of Reasoning in Philosophy":

> To the same natural effects, the same causes must be assigned.

The example he gave was that the light of the Sun and the light of a fire should have the same cause — he knew nothing of nuclear physics. Prudent scientists will be humble enough to admit that there will almost certainly be scientific principles of which *they* are unaware.

To establish a foundation for discussing global warming, it is first necessary to describe the scientific basis for a belief in its validity in terms of the science that we do know.

19.3.1.   *The greenhouse effect*

When a material is described as transparent, it is usually taken to mean that it transmits all or some part of the visible spectrum. Visible radiation is just a small part of the electromagnetic spectrum that, in its totality, stretches from radio waves with wavelengths hundreds of metres or more to γ-rays with wavelengths of 10^{-12} metres or less. We know that brick walls are transparent to radio waves because we can receive radio signals inside a house, but light will not pass through a brick wall. Similarly, a sheet of lead will block out all electromagnetic radiation except γ-rays. The message here is that the definition of transparency should not just be restricted to visible radiation, and that all materials — solids, liquids and gases — will be transparent to some wavelengths and opaque to others.

Any body emits a range of electromagnetic radiation that is characteristic of its temperature. A radiator of a central heating system will emit heat radiation, typically of wavelength 4×10^{-5} m, while the filament of a light bulb, at a temperature of about 2,500 K,[a] emits a great deal of visible radiation in the wavelength range $4-7 \times 10^{-7}$ m. The general rule is that, the higher the temperature is, the shorter the bulk of the radiation being emitted. In the discussion which follows, the radiation of interest is that emitted by the Sun, at an effective temperature of 5,850 K, and by the Earth, at a typical average temperature of 288 K. The distribution of the emitted radiation with wavelength, expressed in terms of the energy per unit wavelength range per unit area per unit time, for sources at these temperatures is shown in Fig. 19.4.

Although it emits a considerable amount of light, most of the radiation from the Sun is heat radiation, while *all* the radiation from the Earth is heat radiation. Another thing to notice is that both curves are scaled to having the maximum of emission at 5,850 K as unity and this shows how small the radiation per unit area is from the Earth, as compared with that from the Sun.

The energy density of solar radiation falling on the Earth per unit time is given by the *solar constant*, $1.366 \, \text{kWm}^{-2}$. This would be the intensity of the radiation falling on the Earth's surface if the Sun were directly overhead and there was no atmosphere. At high latitudes, the intensity is less due to an obliquity factor. However, there *is* an atmosphere — one that is not equally transparent

---

[a]K is the symbol for the Kelvin, or absolute, scale of temperature. On this scale, the temperature of melting ice is 273 K and that of boiling water is 373 K.

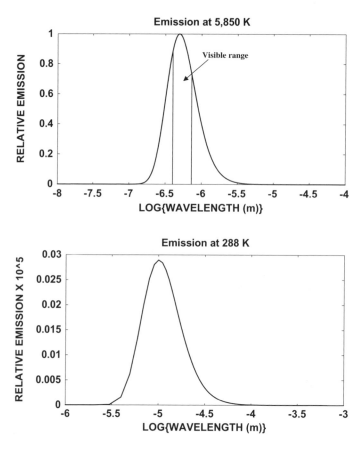

Fig. 19.4. Radiation at solar temperature, 5,850 K and Earth temperature, 288 K.

to all wavelengths, and with all the gaseous components within it providing different contributions to the opacity at different wavelengths.

The overall transmission of the atmosphere in the wavelength range 10^{-7} to 10^{-3} m is shown in Fig. 19.5, together with the positions of the peaks of the transmission curves at 5,850 K and 288 K, as indicated by Fig. 19.4. Comparing the two figures, what will be seen is that, on the whole, the atmosphere is much more transparent to solar radiation than to that from Earth; much of the Earth's radiation is at wavelengths where the atmosphere is completely opaque.

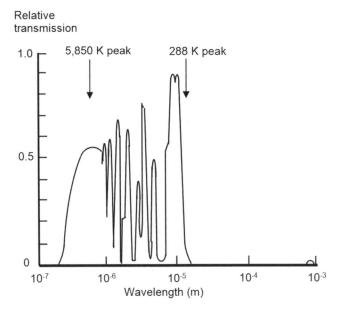

Fig. 19.5.   The variation of the transmission of the atmosphere with wavelength.

The equilibrium temperature reached by the Earth is such that the amount of energy it absorbs from the Sun equals the amount it radiates back into space. The greater the opacity of the atmosphere to the Earth's radiation, the higher the temperature of the Earth will need to be for it to radiate enough energy through the atmosphere to maintain equilibrium. That is the basis of the *greenhouse effect*, so called because the glass of a greenhouse acts like the atmosphere, transmitting solar radiation more or less unhindered, while acting as a barrier to the escape of radiation from the lower temperature interior. Any agency that increases the opacity of the atmosphere to longer wavelength radiation, such as that mainly radiated by the Earth, will increase the equilibrium temperature of the Earth.

It must not be thought that the greenhouse effect is totally harmful. If the atmosphere were completely transparent to all wavelengths, so there was no greenhouse effect, then the mean temperature of the Earth would be about −18°C, making life on Earth much more difficult to sustain.

19.3.2.   *Greenhouse gases*

The regions of opacity in the complicated transmission curve shown in Fig. 19.5, which are the basis of the greenhouse effect, are due to components of the atmosphere known as *greenhouse gases*. It is impossible to ascribe exact contributions to each of these gases, because the regions in which they contribute to opacity overlap to a great extent, but it is possible to identify water vapour (H_2O) as the most important single component. This is one over which humankind has no control; about 70% of the Earth's surface is covered by water and the atmospheric water vapour comes from evaporation from the oceans. The proportion of water vapour in the atmosphere is strongly affected by locality. Over and near large bodies of water, it may constitute up to 4% of the volume of the atmosphere, although paradoxically the Atacama Desert in Chile, next to the Pacific Ocean, is the driest location on Earth; there is virtually no water vapour in the atmosphere, which is why large telescope installations have been built there.

The next most important source of the greenhouse effect is carbon dioxide (CO_2) and here humankind has a strong influence on the carbon dioxide levels in the atmosphere: at present 0.04% by volume. Carbon dioxide is actually an *essential* component of the atmosphere for advanced life on Earth. Photosynthesis in plants extracts carbon dioxide from the atmosphere, combines it with water and produces cellulose — the main structural substance of the plant — plus oxygen. For animate life, plants are the ultimate food source and oxygen is required for respiration. Carbon dioxide enters the atmosphere from many natural sources, for example, the gases contained in volcanic eruptions, and through forest fires that sometimes occur spontaneously during very dry spells. However, since the beginning of the Industrial Revolution in the eighteenth century, the most important input has been the burning of fossil fuels: coal, oil and natural gas. When the Earth first formed, the atmosphere was carbon rich, containing methane (CH_4) and carbon dioxide. Over hundreds of millions of years, organic material, both animate and inanimate, was buried deep underground and converted into other forms — almost pure carbon in the case of coal and hydrocarbons for oil — and the carbon they contained was safely locked away. Now humankind is recovering some of these materials and releasing the carbon, in the form of CO_2, back into the atmosphere.

There are other, less important, greenhouse gases in the atmosphere, notably methane, and nitrous oxide (N_2O) both at very low volume concentrations: 0.00018% for methane and 7×10^{-6}% for nitrous oxide. Methane is actually a much more effective greenhouse gas than carbon dioxide, about 80 times more effective, but its overall effect is smaller because of its low concentration in the atmosphere. There is a great deal of methane locked within the Earth's crust in various forms. It is the dominant component in the natural gas that is exploited as a fuel and is also produced by the decay of organic material. If the mud at the bottom of a pond is stirred, then it will release bubbles of *marsh gas*, which is methane. In large areas of permanently frozen tundra, such as exist over much of northern Russia, methane is locked into the soil — permanently so as long as the soil stays frozen. There is a fear that global warming, initially produced by increased carbon dioxide, may progressively melt the tundra soil, releasing methane, and so provide a positive feedback to the process, leading to runaway warming.

19.3.3.   *The evidence for global warming*

The scientific results displayed in Figs. 19.4 and 19.5 are the results of sound theory, observation and experiment and are unchallengeable. However, by themselves they do not present an unanswerable case for global warming; the Earth is a complex system, with many different components that interact in complicated ways. One thing that can be done is to see how the atmosphere and the Earth's temperature have changed in the recent past; as always, a study of history may give insights into the future.

One of the problems in presenting the temperature of the Earth is that it is highly non-uniform, varying from place to place, and over time during a year as the seasons change. To estimate the change of temperature at a particular location, the average temperature over the year is compared with the average temperature, at the same location, taken over the 30 years 1961–1990. This difference, the *local temperature anomaly*, is then averaged over many widely spaced locations covering the Earth to give a *global temperature anomaly*, which may then be used to estimate long-term trends in global temperature change. Figure 19.6 shows the variation of the global temperature anomaly for the period 1882 to 2005; to smooth out rather abrupt changes from one year to the next, what is plotted is five year averages, i.e., the anomaly shown for 1990 is the average for the five years 1988 to 1992.

It is clear from Fig. 19.6 that there has been a rise in average global temperature in the period shown. The importance of considering the trend over a long period of time is also highlighted by this graph. In the period from 1940 to 1970, it might have been

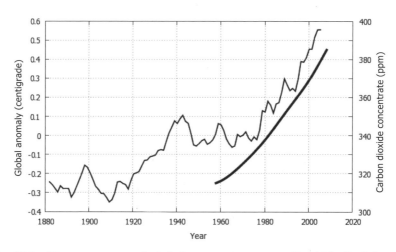

Fig. 19.6. Five-year-averaged global temperature anomalies (thin line) for the years 1882 to 2005 and atmospheric carbon dioxide concentrations (thick line) from 1957 to 2008.

concluded that mean global temperature was stable or even falling, and the period from 1900 to 1910 might have given fears of catastrophic global cooling.

We must now consider the reason for the temperature increase. The first and most obvious phenomenon to consider is the greenhouse effect, in particular the presence of atmospheric carbon dioxide, the only greenhouse gas strongly influenced by humankind. The concentration of carbon dioxide[b] from 1957 to 2008, also shown in Fig. 19.6, seems to be strongly correlated with the temperature increase in the same period.

The dangers of drawing conclusions from observations over a limited period have been mentioned in relation to Fig. 19.6 and, in the context of the age of the Earth, even the whole period covered in that figure might be too short to draw strong conclusions. Fortunately, it is possible to deduce the variations of carbon dioxide concentration and temperature over the last few hundred thousand years, from measurements on ice cores taken in the Antarctic. Antarctic ice is more than three kilometres deep; the snow that falls there is compacted by the weight of subsequent snow, so forming a thick sheet of ice. Consequently, by drilling into the ice and extracting an ice core, it is possible to examine ice formed at an estimated time in the distant past — the time estimate being based on the assumption of a uniform rate of fall of snow. An ice core taken to a depth of over 3.6 kilometres at the Vostok station in Antarctica gave ice samples up to 800,000 years old.

At the time that ice is formed it contains information about the prevailing conditions — for example, pockets of air trapped in the ice give the composition of the atmosphere, including the concentration of carbon dioxide. From the ice it is also possible to find the prevailing temperature. Water is made up of hydrogen and oxygen, both elements consisting of different isotopes. The oxygen in water

---

[b]The concentration is expressed in parts per million (ppm). Sometimes "ppmv" is used to stress that it is the proportion by volume.

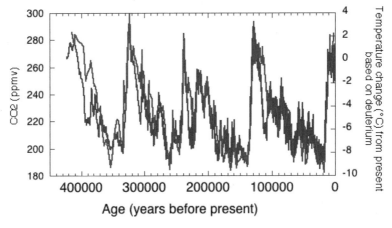

Fig. 19.7. Carbon dioxide concentrations (red) and temperature variation (blue) over the past 400,000 years (NOAA).

contains three isotopes: O16, O17 and O18, all oxygen in the way they behave chemically, but with different atomic masses proportional to the appended numbers. Similarly, hydrogen in water has two isotopes: hydrogen and deuterium, with masses 1 and 2 respectively, on the same scale as the oxygen masses. When evaporation of water takes place, the water molecules containing lighter isotopes are lost more readily, and the rate of evaporation increases with temperature. Consequently, by measuring variations in the isotopic composition of the ice, it is possible to estimate the temperature at the time it was formed. The variations of carbon dioxide concentrations and temperature, as determined in this way, are shown in Fig. 19.7.

The correlation between carbon dioxide concentration and temperature is quite striking, and has been given as unequivocal proof of the hypothesis of global warming.

19.3.4. *The controversy*

Rather curiously, the information contained in Fig. 19.7 has been advanced by the opponents of the human-induced global warming hypothesis as evidence *against* global warming, and the reason they give cannot be dismissed out of hand. Very careful statistical analysis of the data shows that the best fit of the carbon dioxide and temperature profiles is when the carbon dioxide peaks are about 800 years *after* the temperature peaks. The implication that the opponents of global warming draw from this is that temperature variations are due to other causes, and that carbon dioxide variations are due to temperature changes and not vice versa. Indeed, vast amounts of carbon dioxide are present in sea water, and an increase of its temperature would lead to a reduction in the solubility of carbon dioxide and hence its release into the atmosphere.

It is well known that, over long periods of time, the temperature of the Earth has been very variable. This has been due to a number of factors, such as variation in the solar output of energy and periodic changes in the Earth's orbit due to perturbations by other planets. In the past, the Earth has been much colder than at present, and has undergone glacial periods at intervals of between 40,000 and 100,000 years. There have also been occasional periods when the Earth has been much hotter than it is at present. It might be argued from this that the dominant forces that control temperature are beyond human control, and dwarf any human influence. Taken in conjunction with the assumption that carbon dioxide concentrations are driven by temperature, and not the other way round, this is the main argument advanced by those opposing the proposition that recent global warming is human-induced and may give devastating consequences.

19.3.5.   *Can we know for certain?*

We have previously met situations in which the same information has been advanced to support opposite arguments — for example, whether spending more money on a health service that gives better treatment indicates a better medical service or, if at the same time it gives less treatment per unit expenditure, the service has deteriorated. In that case, one can rationally support either side of the argument, depending on one's personal priorities, but the same cannot be true for global warming. The possibilities range from the phenomenon not occurring at all, to it having a powerful effect that will have disastrous consequences for humankind and all other life on Earth. There are intermediate possibilities, such as the phenomenon occurring but with consequences so mild we should not be unduly concerned about them.

The first point to be made is that there is no doubt at all that the greenhouse effect is real and demonstrable. Without it the average temperature on Earth would be about 33°C lower than it is now, and life on Earth would face very different challenges. An even more spectacular example of global warming is presented by the planet Venus, similar in size to the Earth but closer to the Sun. Venus has a 96% carbon dioxide atmosphere with a surface pressure 92 times that on Earth; without its atmosphere the mean temperature of Venus would be 80°C, but the actual value is 460°C, which is hot enough to melt metals such as tin, lead and zinc.

The claim by the opponents of global warming that Fig. 19.7 shows that carbon dioxide levels are controlled by temperature and not vice versa, is not necessarily valid. Carbon dioxide occurs in many locations. Some of it is chemically locked into carbonates, the most prevalent of which is calcium carbonate ($CaCO_3$) which forms the shells of many creatures, the remains of which, over time, form chalk, as seen in the white cliffs of Dover. As a proportion of the total CO_2 on Earth, comparatively little is in the atmosphere; the main reservoir of carbon dioxide is that dissolved in the oceans: some 60 times more than in the atmosphere. Its importance in our present discussion is that, under suitable conditions, it is easily taken up by the oceans or easily released into the atmosphere. The solubility of gases in water is highly temperature dependent, decreasing as the temperature increases. This gives a way of interpreting the long-term data given in Fig. 19.7. A small rise in temperature, initially due to some change in radiation received from the Sun, can trigger a positive feedback mechanism in which a rise of temperature leads to the release of oceanic carbon dioxide, which increases the concentration in the atmosphere, which in turn gives an increased greenhouse effect and a further rise in temperature. If the external heating now reverses to reduce the temperature slightly, this leads to more carbon dioxide dissolving in the sea, a reduction in atmospheric carbon dioxide concentration, less global warming and a lower temperature, with the feedback now working in the reverse direction to reduce the temperature.

The orbit of the Earth around the Sun is affected by other planets, mostly by Jupiter, and leads to a periodic variation, with a period of about 100,000 years. Figure 19.7 indicates a more or less cyclic variation of temperature and carbon dioxide concentration with about this period. A simple computer programme has been used to confirm that a model based on a long-period external variation of solar radiation, plus rates of carbon dioxide production and temperature variation dependent on each other, can reproduce the major features of Fig. 19.7. The parameters used in the programme are as follows:

(i) There was a variation in solar radiation with a period of 100,000 years. The rate of increase of temperature due to this cause alone for the first 90,000 years was 1.55×10^{-5} (14/90,000) K per year, and the decrease in the final 10,000 years was 1.4×10^{-4} (14/10,000) K per year.

(ii) The concentrations of carbon dioxide, C (ppm), and temperature, T (K), giving equilibrium, i.e., such that neither would change due to the other, was taken as:

$$C_{equil} = 110(T - 285)/12 + 180$$

$$\text{giving} \quad T_{equil} = 12(C - 180)/110 + 285.$$

(iii) In a non-equilibrium state the rate of change of temperature is $2 \times 10^{-6}(C - C_{equil})$ K/year and that of carbon dioxide is

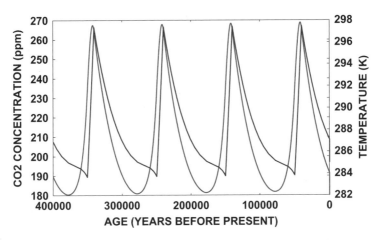

Fig. 19.8. Results from a computer model. CO_2 concentration (red), temperature (blue).

$10^{-3}(T_{t-1000} - T_{equil})$ ppm/year. The quantity T_{t-1000} gives the temperature 1,000 prior to the time at which the carbon dioxide is either released by the oceans or absorbed by them. This allows for the high thermal capacity of the oceans, whose temperature will lag behind that of the atmosphere. This is common knowledge to those brave enough to bathe in the North Sea; even on the hottest summer day it can be a chilling experience.

Running this programme gives the outcome shown in Fig. 19.8. It is a very simple programme and does not include shorter-term fluctuations that are suggested by Fig. 19.7 but it does show a similar range of carbon dioxide concentration and temperature and, importantly, gives a lag of the carbon dioxide concentration peak behind that of temperature. Despite the crudeness of the model, and the fact that the choice of parameters was guided by an attempt to match Fig. 19.7 as closely as possible, it does show that one should not read too much into the lag of carbon dioxide behind temperature. It is not a conclusive argument to refute global warming.

There are many factors, some known and understood and probably others not known, which could affect the efficacy of the global warming process. For example, some of the radiation falling on a planet is reflected back into space by clouds, and hence does not reach its surface. The proportion reflected is called the *albedo* of the planet and is 0.75 for Venus, which is totally blanketed by cloud. For the Earth, it is between 0.30 and 0.35, depending on the amount of cloud cover. If an increase of temperature gave an increase of cloud cover, then less solar radiation would reach the Earth's surface, which would partially compensate for greenhouse gas absorption and reduce the severity of global warming.

In the above simple computer model the trigger for temperature rises and falls was the rise and fall of incident solar radiation, which initiated a positive feedback mechanism involving a linkage between temperature and carbon dioxide concentration. A similar result would come about if the trigger was an imposed alternating rise and fall of carbon dioxide concentrations, although a natural mechanism for achieving this is hard to envisage. However, there is a mechanism for steadily increasing the CO_2 concentration — the burning of fossil fuels — and, in the absence of compensating factors, this would lead to an inexorable rise in the temperature of the Earth.

19.3.6.   *Better safe than sorry*

To summarize: global warming is undoubtedly a valid phenomenon, with a probability very close to one. The recent range of carbon dioxide concentrations, as shown in Fig. 19.6, is entirely above that given in Fig. 19.7, and is increasing. There are other factors of uncertain importance — for example, the release of methane from frozen tundra in a direction of increasing temperature, and extra cloud formation giving an effect in the opposite direction. The observations recorded in Fig. 19.7 have been related to variations in the Earth's orbit, without any input from human activity, but it seems almost certain that as the carbon dioxide concentration increases, so average temperatures will also increase. The debate should not be about whether or not global warming will occur — it almost certainly will — but the extent of the warming, whether or not there will be braking mechanisms to limit its effect, and what the likely consequences are for climatic change and its effect on humankind.

It could be argued strongly that even if, in the light of present knowledge, the probability of severe global warming was very small, which it is certainly not, the consequences of allowing it to happen would be so severe that one should not gamble on it not happening. The worst outcome if we go all-out to adopt a low-carbon economy is that we may expend resources unnecessarily and become a little poorer — although many argue, to the contrary, that if we make the investment in renewable energy, we would actually be better off in the longer term. The world should be embarking on an *energy revolution*, which would be every bit as important as the Industrial Revolution that began to transform the world in the eighteenth century. Any country that seized the initiative to spearhead this revolution would gain great economic benefit, as did Great Britain when it became the "workshop of the world" during the Industrial Revolution.

The worst that can happen if we do nothing is of a different order completely. Ice melting at the poles would severely affect the global climate and lead to a rise in sea levels that would remove productive land required to feed an ever-increasing world population. The production of food could also be affected by floods and droughts on a much greater scale than those now experienced. There would be social pressures, and even conflicts, brought about by migrating populations and the competition for scarce resources.

Such scientific evidence as we are sure about points towards a global catastrophe if we do nothing. Factors that may reduce the effect of global warming are at the level of speculation; they may exist but the probabilities are low. On the basis of this evidence, should we gamble on the effects of global warming? The stakes are high.

19.4.   Nuclear Power Generation

There are many areas of public life in which science plays a role: weather forecasting, global warming, nuclear power generation, the building of dams and tidal barriers, the development of new drugs and the growing of genetically-modified crops, for example. The public expects definitive answers and definitive solutions concerning safety and other matters related to these topics, and the problem is that such answers are impossible to provide. In the 1950s, the drug thalidomide was tested and found to be an effective antidepressant, painkiller and antiemetic, and it was prescribed to pregnant women as an antidote to morning sickness. Subsequently, there was a significantly higher proportion of babies born with stunted arms or legs and, eventually, it was found to be due to thalidomide. It had been widely believed that the transfer of harmful agents across the placenta, the lining of the womb within which the foetus develops and that prevents the mixing of the mother's and baby's blood, was not possible — but it was. It was a tragedy, but one that could not have been foreseen, and one of a type that may occur again as long as new drugs are being developed. Can it be said that any new drug is completely safe, given that side effects may take years to manifest themselves? The answer is no, but the alternative of not developing new drugs is not a realistic option.

One topical question that may be asked is: "How safe is nuclear power generation?" There have been two major disasters at nuclear power stations: at Chernobyl in the Ukraine in April 1986 (when it was part of the Soviet Union), and in Fukushima, Japan, in March 2011. The Chernobyl disaster was caused by a combination of poor reactor design — fortunately, it was the only reactor of its kind — and human error. A test of the reactor was being carried out at a time when it was in a near unstable condition and the test tipped it over into instability. It exploded and released into the atmosphere radioactive materials, which were mostly deposited in parts of northern Europe. In the case of Fukushima, the trigger for the failure was a massive earthquake, the most powerful in Japan since records began. The four reactors which comprised the power station were about 70 km from the epicentre of the earthquake and were not seriously affected by the shock but, as part of built-in safety precautions, they were automatically shut down. Subsequently, emergency power generators were used to power the equipment that cooled the reactor cores. However, a subsequent tsunami, with a wall of water almost 38 m in height, disabled the emergency power generators, allowing fuel rods in the reactors to melt. Explosions, probably due to hydrogen gas, damaged the fabric of the reactors, causing further problems concerning the leakage of radioactive material.

On the basis of these two cases can we say that nuclear reactors are unsafe? Chernobyl seems to be a unique case, and is unlikely to be repeated. The reactor was of poor design and the people running it seemed to have been poorly trained. Fukushima is different. The problem there consisted of not anticipating a series of events that turned a safe reactor into a nuclear disaster. This anticipation of a variety of circumstances that may occur, and may have to be dealt with, is known to commercial and military organizations as *contingency planning*, and the challenge is to envisage all the contingencies that may occur — however outlandish. There is a story, perhaps apocryphal, that after the 1967 Arab–Israel war, an Israeli general, being interviewed by journalists, was asked what contingency plans they had for the campaign they had just fought. The reply was that they had contingency plans for everything — including fighting above the Arctic Circle. The story may not be true, but illustrates the point that you must consider what you would do, not only in likely situations that may occur, but also in very unlikely, but possible, situations. An earthquake followed by a tsunami is a well-known phenomenon that had occurred many times previously, but the implications of such a double event had not been allowed for by those running the Fukushima power station.

There are minor incidents that occur all the time in nuclear power stations, as occur in all major industrial plants, but, set against the number of nuclear power stations that now exist in the world, the occurrence of major incidents is at a level that might be expected in, say, the chemical industry. The difference with nuclear disasters is that the radioactive materials that leak out, and that are harmful to life, have half-lives of many decades, meaning that a considerable area around the stricken reactor is unsafe for habitation for periods of the order of centuries.

The overall situation is that, with all the fail-safe features of nuclear reactors and with the best proven safety designs, at the end of the day there are fallible human beings involved in running them, and events may occur that are so unlikely that they cannot be planned for. For example, it is not impossible, but extremely unlikely, that a small asteroid could fall on a nuclear power station causing critical damage, but it is not so unlikely that a terrorist may decide to highjack an aircraft and crash it onto the power station. If a policy is followed of only building nuclear power stations in seismically quiet and remote areas, and if there is a Fukushima-magnitude incident every 25 years on average, then should we go ahead with nuclear power generation as a remedy for global warming? That is a decision to be made by informed citizens through their governments, not by scientists alone. Scientists and engineers can use their skills to minimize the risks but they cannot remove them entirely; what they can do is to quantify the risk as best they can, to guide those making the ultimate decisions.

The same considerations apply in other areas where science and engineering impinge on everyday life. Scientists cannot give yes or no answers to all questions. If we want to know if Monday will be free of rain, then we must accept the answer that there is a 30% chance of precipitation and decide for ourselves whether or not to take the risk that it will not rain. Unfortunately, there is only a small proportion of the population at large that will accept that scientists cannot provide definitive answers. They want a yes or no response to the question of whether it will be dry on Monday and will assume that the scientist is incompetent if the answer is incorrect. On the other side of the coin, there are scientists who are so keen to promote a particular point of view, for example, on global warming, that they sometimes present information in a distorted way to strengthen their case. There is never any excuse for this. The role of scientists is to determine the facts as best they can, and to present them to the public objectively and without distortion in order that governments, representing society, can make sensible informed decisions. Scientists can, of course, express opinions and give advice — that is expected of them — but they must be objective opinions and advice, free of bias.

19.5. Decision-Making Without Hard Data

At the beginning of Chapter 13 we gave a quote from Bismarck: "Politics is not an exact science". In view of what has been written in this chapter, we may make the paradoxical statement that "Science is not an exact science", despite a popular view to the contrary. Scientists must accompany any advice they give with some idea, however crude, of the likelihood that the science on which it is based is sound.

In Section 18.5 a mathematical approach was suggested to decide whether the invasion of Iraq in 2003 was a good or bad decision. To decide on the issue, it was necessary to estimate the probabilities of a large number of outcomes, such as the occurrence of sectarian strife, with the uncertainty that unpredicted outcomes might occur, and that all probabilities were pure guesswork in any case. The situation is similar, but somewhat worse, for making decisions on those scientific matters that affect society at large. For all that scientists do know, the study of the history of science suggests that there is still much that they do not know. This means that there are an unknown number of outcomes that cannot be predicted, which leaves gaps in any probability tree we might wish to construct. There is also uncertainty in any numerical probabilities assigned to various possible outcomes, although history may give guidance. Clinical trials on new drugs have been going on for many years, and the proportion of times that drugs are given clearance but turn out to be harmful can probably be assessed reasonably well. On the other hand, the frequency of major accidents with nuclear reactors, given as 1 per 25 years in the previous section, may be wildly out in either direction. Major reactor accidents are rare events and it will take a long time, perhaps centuries, to get a reasonable assessment of frequency.

There are other factors that must be taken into account when considering issues that could conceivably affect the survival of humanity and other life forms. If an action gives a 1 in 1 million chance that it could go wrong and destroy the planet, but if it goes right will give some benefit to humanity, then should we take that action? Never in the past have we been faced with such a decision — although global warming now gives us a similar question to answer, with the difference that the probability that global warming is occurring, and will do immense harm, is high, not low. Other decisions we take are important without the fate of the whole of humankind being an issue. We know that genetically-engineered plants can give benefit, such as increased yields that help to feed a growing world population. There is a small chance that they could be harmful — perhaps affecting human genes in some adverse way; while there is no evidence that it has ever occurred, we must accept the possibility that it could. Scientists would point out in defence of genetic engineering that by selective breeding we have been carrying out a form of genetic engineering for a very long time, albeit rather slowly and inefficiently, with no known adverse consequences.

The advance of civilization has conferred many benefits on humankind but has also introduced new hazards. It is likely that humanity could have survived indefinitely with the Black Death (giving a chaotically fluctuating population), bows and arrows, and burning wood for fuel. It is not so certain that we can survive indefinitely with antibiotics (giving an ever-growing population), nuclear weapons and fossil fuels. However, there is no turning back.

Problem 19

19.1 From Fig. 19.3 it is estimated that between 1980 and 2000 the concentration of CO_2 in the atmosphere increased by 37 ppm and the global anomaly increased by 0.4°C. Assuming the same rate of increase, what will be the CO_2 concentration and global anomaly in the years 2100 and 2200?

   If the world abruptly halved emissions of CO_2 in 2020, then what would be the CO_2 concentration and global anomaly in the years 2100 and 2200?

   (*Hint*: You will need to make estimates of values from Fig. 19.3.)

Chapter 20

The Pensions Problem

The days of our years are threescore years and ten; and if by reason of strength they be fourscore years, yet is their strength labour and sorrow, for it is soon cut off, and we fly away (Psalms 90:10)

20.1. A History of Life Expectancy

A baby born in Britain in the fourteenth century faced a very perilous future. There would be about a one-in-four chance of dying by the age of five from some disease or other, which was not surprising since people in those days had no sanitation and, especially in cities, lived and worked in an environment almost ankle-deep in human excrement. Those who survived through infancy probably acquired some immunity from the common diseases prevalent at the time, although there would still be the peril of intermittent outbreaks of the Black Death (bubonic plague) to confront; an outbreak in the middle of the fourteenth century is estimated to have killed at least one-third of the population of Europe. Of course, there were some who avoided or survived these threats to life, and managed to live into their fifties and beyond, but the life expectancy, or average lifespan, of someone born in that century was probably about 35 years.

Life expectancy did not change by much in the next three or four hundred years, although by the beginning of the nineteenth century it had crept up to about 40. Then, in the nineteenth century, there were several events, the consequences of which gave a dramatic improvement in life expectancy. All these events led to a better understanding of the causes of disease and of the steps required to reduce the incidence of disease. The Crimean War (1853–1856), in which France and Britain confronted Russia, was noteworthy for the very high mortality rate of British soldiers. Newspaper reports described the dire conditions faced by wounded British soldiers in hospital, with inadequate medical supplies and filthy wards. Many more were dying of diseases acquired while in hospital than were dying of wounds. At the request of the British government, Florence Nightingale (1820–1910) went with a team of nurses to the military hospital at Scutari in Turkey. She instituted a regime of absolute cleanliness and strict hygiene, and the mortality rate was drastically reduced. The lessons learned in Scutari were subsequently applied to all hospitals.

During the nineteenth century there were several serious cholera outbreaks in London, with particularly bad ones occurring in the 1830s and 1840s. In 1854 there was a serious outbreak in the Soho area of London. In those days, many people got their water from public water pumps situated in the street. A London doctor, John Snow (1813–1858), through some careful detective work, found that all the victims had used one particular water pump, and so he identified contaminated water from that pump as the cause of the disease. Prior to this discovery it was generally believed that diseases were caused by "bad air" or *miasmas*. The final abandonment of that theory was due to the work of the French scientist Louis Pasteur (1822–1895), who showed that it was germs entering the body by some means or other that caused disease, and he imaged some of these disease-causing agencies using a microscope.

Another nineteenth-century landmark event involving sanitation was the so-called "Great Stink" of 1858, when the whole of central London was affected by foul smells emanating from the River Thames due to untreated sewage pouring into the river. At that time most human waste went into large cesspits; this mixture of human excrement and urine, called *night soil* because the cesspits were emptied at night by collectors at intervals of a few weeks, was sometimes used as a fertilizer — and still is in some Third

World countries — with obvious risks to health. In the nineteenth century there was an increase in the number of flush toilets; the extra water they used was flooding the cesspits, with the overflow ending up either in the Thames or polluting drinking water supplies. Since the Houses of Parliament were beside the Thames, politicians were particularly affected, and they initiated an enquiry into how to solve the problem. The solution was proposed and implemented by Joseph Bazalgette (1819–1891), the Chief Engineer of the Metropolitan Board of Works, and a mammoth project of sewer building was undertaken under his direction. Hundreds of miles of main and subsidiary sewers were completed in the remarkably short time of nine years. Not only did this solve the problem of the bad smell, but also effectively separated sewage from the drinking water supply, and so greatly improved the overall health of Londoners.

As a result of these events, and also improvements in diet (for example, the widespread availability of potatoes which became an important component in the diet of most people), by the beginning of the twentieth century life expectancy had increased to 47. This was an increase of seven years in a century.

If the increase in life expectancy was substantial in the nineteenth century, then the increase in the twentieth century can only be described as spectacular — not only in Britain but worldwide, especially in more affluent countries. This was due to improvements in medical practice (especially the introduction of antibiotics and drugs that cured previously fatal diseases), better access to medical services, better diet and improved personal hygiene. At the beginning of the century few homes had bathrooms and inside toilets; by the end of the century virtually all homes had those facilities. In Britain during the course of the twentieth century infant mortality fell from 140 to less than 6 per 1,000 births and life expectancy rose to almost 79 years. Similar patterns of improvement were experienced in many countries; Table 20.1 gives an extract from tables published by the United Nations covering the years 2005–2010 for its 194 constituent members. In general, every tenth entry is given in the table, plus the

Table 20.1. Life expectancy (2005–2010) for a selection of member countries of the United Nations (UK Office for National Statistics).

Position	Country	Average	Men	Women	Position	Country	Average	Men	Women
1	Japan	82.6	78.0	86.1	110	Indonesia	70.7	68.7	72.7
11	Canada	80.7	78.3	82.9	111	Thailand	70.6	66.5	75.0
21	UK	79.4	77.2	81.6	121	Maldives	68.5	67.6	69.5
31	Luxembourg	78.7	75.7	81.6	131	Western Sahara	65.9	64.3	68.1
36	USA	78.3	75.6	80.8	139	India	64.7	63.2	66.4
41	Kuwait	77.6	76.0	79.9	141	Mauritania	64.2	62.4	66.0
51	French Guiana	75.9	72.5	79.9	151	Ghana	60.0	59.6	60.5
61	Ecuador	75.0	72.1	78.0	161	Gabon	56.7	56.4	57.1
71	Serbia	74.0	71.7	76.3	171	Burkina Faso	52.3	50.7	53.8
81	China	73.0	71.3	74.8	181	Somalia	48.2	46.9	49.4
91	Sri Lanka	72.4	68.8	76.2	191	Lesotho	42.6	42.9	42.3
101	Philippines	71.7	69.5	73.9	194	Swaziland	39.6	39.8	39.4

figures for the world's four most populous countries: China, India, the United States and Indonesia, and also the bottom marker of the table: Swaziland. For most countries the life expectancy of women is higher than for men, so, as well as an average, the figures are given separately for men and women.

For countries at the top of the list the figures will reflect the quality and availability of medical services. However, the national figures conceal a great deal of variation within countries. For example, in prosperous Kensington and Chelsea in London, the life expectancy is 84.4 years for men and 89.0 years for women, well above the national average. However, in Calton, a poor district in East Glasgow (Scotland), the life expectancy for men is just 54 years, comparable to that in some of the poorest states in Africa. The reason for this dreadful statistic is well understood: excessive smoking (twice the average national rate), excessive alcohol consumption and a very poor diet. Nearly all the countries at the bottom of the list are in sub-Saharan Africa and, apart from the adverse influence of poverty on life expectancy, many of them also have a massive problem with HIV/AIDS.

There is no reason to suppose that life expectancy will not continue to increase, at least in the more affluent countries in which medical advances are available to most people. It has been predicted that the number of centenarians will steadily rise, and the proportion of the population beyond the age when they would be expected to work will also increase; it has been predicted that, in Britain, one quarter of girls under the age of 16 in 2011 will live to be more than 100 (BBC Magazine, 9 October 2009, commenting on a report by the UK Office for National Statistics). While longevity is generally regarded as a benefit to humankind, it does throw up severe social problems for society at large.

20.2. Supporting the Aged, and Others

In mediaeval times the proportion of people who survived to an age beyond their capacity to work was much smaller than now, and those that did so, plus those disabled and unable to work, would depend on family support, begging or the alms offered by the Church.

One effect of the Black Death pandemic in the middle of the fourteenth century was that there was a shortage of labour in England. To prevent people from moving around the country, offering their labour for ever higher wages, thus leading to spiralling inflation, a law was passed that fixed wages and made vagrancy a criminal act, which effectively restrained workers to remain within their parishes. The clergy, who were often in contact with plague victims through the administering of the last rites, had a particularly high death rate; in many monasteries more than 50% of the priests perished. Eventually there were not enough priests to cope with the demands of the dying and, by special dispensation from the Pope, the dying could make last rite confessions to almost anyone and receive absolution. Another consequence of the shortage of priests was that they were no longer able to adequately perform their function of supporting the poor so, to compensate for this loss of church charity, the state took responsibility for looking after the poor — those who for any reason could not work to support themselves — and appointed a prominent citizen in each parish to administer the charity.

At the beginning of the seventeenth century a new mechanism for dealing with the poor was set up. Workhouses were established that gave accommodation and food to those in need: the old, infirm and those unable to find employment; the last named group, those that could work but were unemployed, were made to work hard, doing mostly menial tasks, in return for what they received. During the nineteenth century, the Industrial Revolution, and the consequent replacement of people by machines, meant that unemployment grew, and the expense of the workhouse system imposed a heavy burden on the state. At the beginning of the twentieth century infant mortality rates were still very high, but life expectancy had risen to 47 and a greater proportion of people were living beyond the age at which they could be expected to work; at that time a large proportion of the workforce was employed in heavy manual labour, for example, coal mining. These older people were unable to support themselves and were often without family support. Some new way of supporting this growing army of elderly people was required, and it came in 1908 in the form of the Old Age Pensions Act, introduced by the Liberal politician David Lloyd-George. This was a non-contributory scheme, funded out of general taxation, which gave 5 shillings (25 p) per week to a single pensioner and 7 shillings and sixpence (37½ p) to a married couple. The pension was kept low to encourage workers to make their own provision for retirement, and it was means-tested so that those with an annual income of more than £21 10s (£21.50) were ineligible to receive it. Those that did receive it had to pass a test that indicated they were of "good character"!

The idea of giving a pension to those who reach a retirement age was not entirely new. In 1670 the Royal Navy introduced pensions for retired naval officers — although other ranks had to fend for themselves at that time. Throughout the twentieth century, and in particular after the Second World War, with the publication of the Beveridge Report by the Liberal economist William Beveridge, a welfare state was established in the United Kingdom which gave support for the unemployed and those disabled and unable to work; a contributory state pension for those who reached retirement age; and a National Health Service, available to all and free at the point of use. The original Lloyd-George scheme set the retirement age at 70, but this was eventually reduced to 65 for men and 60 for women. The basic state pension in 2011 was approximately 20% of the average salary for that year, but an individual's state pension could be augmented by making extra contributions when he or she was working. However, only a minority depend on the state pension alone and most retired people have an income from their company or employer pension schemes or from private pension schemes if they were self-employed. For example, in a typical scheme covering manual workers in the National Health Service, the worker contributes 6% of his or her salary and the employer 14% towards the individual's NHS pension.

The United Kingdom state pension has the form of a compulsory insurance scheme. Everyone contributes, although many will die before reaching retirement age and so will reap no benefit from their contributions. At the other extreme, if someone survives to very old age, then the total benefit they receive will far exceed the contributions they made. This is the very essence of any insurance arrangement — although someone insuring against a mishap, such as a motor-vehicle accident, would, by-and-large, hope to give more than he receives! In a well-balanced state pension scheme the losses due to those that live to a great age will be compensated by the gains from those that die earlier. Clearly, this balance between those that receive more than they contribute and those that receive less, is critically dependent on life expectancy that, as we have seen, is on a rising trend.

20.3.   The Age Structure of Populations

It is clear that, although having a high proportion of the population living to extreme old age might be regarded as socially desirable, it creates severe problems for pension funds. Looking at the problem in terms of what is happening in a particular year, for the pension fund to be viable, the total contributions of those paying into the fund has to be equal to, or greater than, the total being paid to the recipients of pensions. This will depend on the age structure of the population (the proportion of people there are at various ages), the contributions to the fund, and the amount paid to pensioners.

As an example of age structure, Table 20.2 gives the percentage of men and women in various age ranges in the United Kingdom in 2002. Although the table gives a complete account of the age structure, it is not easy to visualize the information it contains. For this reason the same information is normally given in the form of a *population pyramid*, which is shown in Fig. 20.1 for the data in Table 20.2.

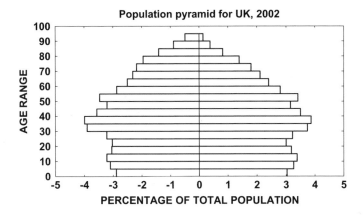

Fig. 20.1.   A population pyramid for the UK in 2002 (left, women; right, men).

Table 20.2.   The age structure for men and women in the United Kingdom in 2002 (UK Office for National Statistics).

Age range	Men (%)	Women (%)	Age range	Men (%)	Women (%)
0–5	3.03	2.89	50–55	3.41	3.46
5–10	3.26	3.10	55–60	2.80	2.87
10–15	3.38	3.22	60–65	2.40	2.50
15–20	3.18	3.05	65–70	2.11	2.31
20–25	3.00	3.03	70–75	1.80	2.18
25–30	3.22	3.22	75–80	1.39	1.95
30–35	3.74	3.90	80–85	0.82	1.41
35–40	3.87	3.99	85–90	0.38	0.89
40–45	3.50	3.56	>90	0.14	0.49
45–50	3.15	3.21			

To the experienced eye this diagram gives a great deal of information, for example, the narrowing of the pyramid for lower age groups tells of a declining birth rate. By contrast, Fig. 20.2 shows the population pyramid for Mexico, a country with a high birth rate and a soaring population: increasing from 15 million in 1900 to more than 110 million in 2010. From the diagram it can be seen that the rate of increase of population has slowed down in recent years, but was still about 1% per year in 2010.

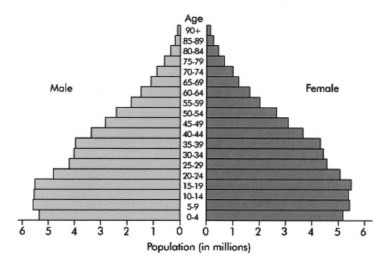

Fig. 20.2. A population pyramid for Mexico in 2010 (US Census Bureau).

Population age-structures are dynamic, and change in ways that are not always predictable. There are some general rules that apply, for example, with increasing prosperity the birth rate tends to decline; in many Third World countries a large family was a form of insurance against destitution in old age, since the custom was for extended families to live together, with the young supporting the old. Even in developed countries there are changes in the age structure with time, with one factor being immigration that tends to bring in younger people from areas where birth rates are higher. A pension fund has to cope with changes of population structure that occur over several decades: 20-year olds paying their pension contributions in 2010 will receive their pension in 2055 or later. Many pension funds have experienced problems, because they have not predicted the future correctly, and find themselves in deficit.

20.4.   The Arithmetic of Pensions

The pensions industry employs large numbers of people, including statisticians known as *actuaries*. Their function is to predict future trends of all the relevant factors that affect pensions, for example, changes in population statistics and the economy, and even to make contingency plans for events that may occur at some time in the future, for example, changes of government policy. They prepare *actuarial tables*, giving mortality rates within a population, and try to predict how these will change with time. Actuaries are highly paid for their mathematical and statistical skills and, clearly, we cannot expect to analyse the problems of the pensions industry bringing in the full range of factors that these experts would deploy. Instead we will take an idealized and a very simple situation that illustrates the basic problem of the pensions industry: balancing what is paid out to pensioners with what is paid into the fund on behalf of those still working.

We shall consider an employer with a workforce between the ages of 20 and 65, with a gender–age structure as shown in Table 20.2, and whose retirees also have that gender–age structure. The average annual income of this workforce is I and the combined employee-plus-employer contribution to the pension fund is a fraction f of the employee's income. If we now add all the percentages in Table 20.2 for men and women between the ages of 20 and 65 (taken as the retirement age for women as well as men), the payment into the fund is proportional to that sum of percentages. This gives total payments to the fund:

$$T = 58.83CfI, \tag{20.1}$$

where C is a constant such that $58.83C$ is the number of people paying into the fund.

The number of recipients of a pension is proportional to all the percentages in Table 20.2 from the age of 65 upwards, and is $15.87C$. We now assume that, because of annual salary increments, the average of the salaries on which the pensions are based is higher than the average salaries of those working. Let this average salary be A, with the pension a fraction p of that salary. The amount paid out in pensions is then:

$$P = 15.87CpA. \tag{20.2}$$

In many pension schemes it is customary to pay the retiree a lump sum which is some fraction of A. Let us take it as equal to A and paid to those aged 65. We take the percentage of people reaching the age of 65 in any one year as one-tenth of those aged between 60 and 70, giving a number of such individuals as $0.93C$. Hence the money paid out each year as a lump sum is:

$$L = 0.93CA. \tag{20.3}$$

Finally, we take into account the administrative expenses of the fund as a fraction s of the total income. Now for the fund not to be in deficit we must have:

$$(1 - s)T \geq P + L. \tag{20.4}$$

Now we consider the following parameters, ignoring the units of currency:

$$I = 20,000; \quad f = 0.2; \quad A = 30,000; \quad p = 0.5; \quad s = 0.05.$$

Inserting these numbers we find the left-hand side of (20.4), the total amount available for paying pensions, is $(1 - s)T = 223,554C$, while the amount paid in pensions is $P = 238,050C$ and in lump sums is $L = 27,900C$. Under these conditions the total outgoings are $265,950C$, so the pension scheme is clearly non-viable. To make the scheme viable it would be necessary to do one, or a combination, of the following:

(i) increase the contributions of employee, employer or both.
(ii) decrease the pension as a fraction of A.
(iii) increase the age of retirement.

Option (i) alone would require the total contribution fraction, f, to be increased to:

$$f = \frac{P + L}{58.83(1 - s)IC} = \frac{265{,}950}{58.83 \times 0.95 \times 20{,}000} = 0.238.$$

Option (ii) alone requires p to be reduced to:

$$p = \frac{(1 - s)T - L}{15.87CA} = \frac{0.95 \times 235{,}320 - 27{,}900}{15.87 \times 30{,}000} = 0.411.$$

Option (iii) is the one proposed by the United Kingdom government in 2011; their proposal is to raise the qualifying age for the state pension by stages to 68. From the point of view of the pension fund it has the double advantage of increasing the number of years of contribution and reducing the numbers of years of receiving a pension. For our present example, increasing the retirement age to 67 gives:

$$T = 60.60CfI, \quad P = 14.10CpA \quad \text{and} \quad L = 0.88CA,$$

where the percentage of those aged between 65 and 67 has been taken as two-fifths of those in the range 65–70 and the percentage reaching retirement is one-fifth of those in the same range. Now the annual income to the fund is:

$$(1 - s)T = 60.60(1 - s)CfI = 230{,}280C$$

and the total paid out is:

$$P + L = (14.10 \times 0.5 \times 30{,}000 + 0.88 \times 30{,}000)C = 237{,}900C.$$

The fund is still not quite in balance but a small change in contribution would make it so.

While our description of a pension scheme has been greatly simplified, it does illustrate the basic problems due to dramatic increases in life expectancy. All three options to bring schemes into balance are to the detriment of the members of the scheme. Both employers and employees do not wish to increase their contributions, which will reduce profits and take-home pay, respectively. Reducing the final pension as a proportion of final income represents a reduction in the standard of living of the pensioner, which is again unwelcome. The acceptability of the third option really depends on the individual and the nature of his or her employment. For many professional occupations work is fulfilling and interesting, and spending a couple of years extra at work may be acceptable; indeed for some it may even be welcome. On the other hand, for those doing strenuous manual work it may not be practicable for them to work the extra years.

The problem will become more severe as life expectancy continues to increase. Eventually, it may be necessary to accept that, if one wishes to have an acceptable standard of living in old age, then a reduction in one's standard of living when working is the price that will have to be paid.

Problem 20

20.1 The managers of the pension fund described in Section 20.4 decide to increase total contributions to 0.22 of salary, reduce pensions to $0.475\,A$ and increase the retirement age to 66. After these changes, by how much will the pension fund be in surplus or deficit each year? Your answer will include the constant C.

Solutions to Problems

Problems 1

1.1 The words showing lack of certainty are underlined.

> *Low pressure is <u>expected</u> to affect northern and western parts of the UK throughout the period. There is <u>a risk of some</u> showery rain over south-eastern parts over the first weekend but otherwise much of eastern England and <u>possibly</u> eastern Scotland <u>should be</u> fine. More central and western parts of the UK are <u>likely</u> to be <u>rather</u> unsettled with showers and <u>some</u> spells of rain <u>at times</u>, along with <u>some</u> periods of strong winds too. However, with a southerly airflow dominating, rather warm conditions are <u>expected</u> with warm weather in any sunshine in the east.*

1.2 Since all numbers are equally likely to occur the probability is: $\frac{1}{12}$.

1.3

Most likely   Least likely

1.4 The fraction of patients who died in the epidemic is $\frac{123}{4,205} = 0.029$. This is the probability that a given patient with the disease will die.

Problems 2

2.1 There are five sides with a face value less than 6 and getting these are mutually exclusive events. Hence the probability of getting a number less than 6 is:

$$P_{1\,to\,5} = \frac{1}{12} + \frac{1}{12} + \frac{1}{12} + \frac{1}{12} + \frac{1}{12} = \frac{5}{12}.$$

2.2 The probability of picking a particular ace is $\frac{1}{52}$. There are 4 aces and getting them are mutually exclusive events. Hence the probability of getting an ace is:

$$P_{ace} = \frac{1}{52} + \frac{1}{52} + \frac{1}{52} + \frac{1}{52} = \frac{1}{13}.$$

2.3 The probabilities that the dodecahedron and die will give a 5 are $\frac{1}{12}$ and $\frac{1}{6}$, respectively. Since getting the 5 s are independent events, the probability of getting the two 5 s is:

$$P_{5+5} = \frac{1}{12} \times \frac{1}{6} = \frac{1}{72}.$$

2.4 From the argument given in 2.2, the probability of picking a jack is $\frac{1}{13}$. For each coin, the probability of getting a head is ½. Since all the events are independent, the total probability is:

$$P_{J+4H} = \frac{1}{13} \times \frac{1}{2} \times \frac{1}{2} \times \frac{1}{2} \times \frac{1}{2} = \frac{1}{104}$$

smaller than that found in 2.3.

2.5 From 2.3 the probability of getting any particular combination of numbers is $\frac{1}{72}$. The combinations that can give a total of 6 are: $1+5, 2+4, 3+3, 4+2$ and $5+1$. These combinations are mutually exclusive, so the probability of getting a sum of 6 is:

$$P_{sum=6} = 5 \times \frac{1}{72} = \frac{5}{72}.$$

2.6 The probability that a particular gene will be:

$$f \text{ is } P_f = \frac{1}{41} \text{ and that it will be } F \text{ is} : P_F = \frac{40}{41}.$$

Note that the sum of the two probabilities is 1 (the gene has to be one or other of the mutually exclusive possibilities) and that the ratio is 1:40 as given.

(i) To have the disease, each gene of the pair possessed by the individual must be f. Hence the probability of having the disease is:

$$P_{ff} = \frac{1}{41} \times \frac{1}{41} = \frac{1}{1,681}.$$

(ii) To be a carrier the gene pair must be (f, F) or (F, f), the combinations being mutually exclusive. Hence the probability of being a carrier is:

$$P_{carrier} = \frac{1}{41} \times \frac{40}{41} + \frac{40}{41} \times \frac{1}{41} = \frac{80}{1,681} = 0.0476.$$

Problems 3

3.1 We calculate the summation on the right-hand side of (3.3) for the three races.

2.30 pm $Sum = \frac{1}{2} + \frac{1}{5} + \frac{1}{7} + \frac{1}{9} + \frac{1}{13} + \frac{1}{19} = 1.0835$

3.15 pm $Sum = \frac{1}{3} + \frac{1}{4} + \frac{1}{4} + \frac{1}{5} + \frac{1}{7} + \frac{1}{9} = 1.2873$

3.45 pm $Sum = \frac{1}{3} + \frac{1}{4} + \frac{1}{5} + \frac{1}{9} + \frac{1}{13} = 0.9714$

The third race does not satisfy the golden rule. The sum is the greatest for the second race and hence is the best from the bookmaker's point of view (and worst for the punters' chances of winning!).

Problems 4

4.1 The number of different orders is $5! = 5 \times 4 \times 3 \times 2 \times 1 = 120$.

4.2 The number of ways is $^{10}C_4 = \frac{10!}{6!4!} = 210$.

4.3 The number of ways of selecting three balls, taking the order into account, is $6 \times 5 \times 3 = 90$. One of those ways is red, green and blue, in that order, so the probability is $\frac{1}{90}$. The ways of selecting three balls without regard to order are:

$$^{6}C_3 = \frac{6!}{3!3!} = 20.$$

One of those non-ordered selections is red + green + blue, so the probability of picking that is: $\frac{1}{20} = 0.05$.

4.4 The number of ways of picking 4 numbers from 20 is $^{20}C_4 = \frac{20 \times 19 \times 18 \times 17}{4 \times 3 \times 2 \times 1} = 4845$. There is one way to get the first prize, so the probability is: $\frac{1}{4,845}$.

The number of different selections of 3 correct numbers from 4 is $^{4}C_3 = 4$, so the number of different ways of getting a second prize, with 3 correct numbers plus the bonus ball is 4. The probability of getting a second prize is thus: $\frac{4}{4,845}$.

A third prize is won, with 3 correct numbers and 1 of the 15 numbers that is neither a correct number nor the bonus ball number. The number of ways of obtaining this is $4 \times 15 = 60$, so the probability of winning a third prize is: $\frac{60}{4,845} = \frac{4}{323}$.

Problems 5

5.1 (i) The number of ways they can have birthdays is 100×100. The number of ways they can have different birthdays is 100×99.
The probability they have different birthdays is: $\frac{100 \times 99}{100 \times 100} = 0.99$.

(ii) The number of ways they can have birthdays is $100 \times 100 \times 100 \times 100$.

The number of ways they can have different birthdays is $100 \times 99 \times 98 \times 97$.

The probability they have different birthdays is: $\frac{100 \times 99 \times 98 \times 97}{100 \times 100 \times 100 \times 100} = 0.9411$.

(iii) We calculate the series of terms:

$$\frac{100}{100} \times \frac{99}{100} \times \frac{98}{100} \times \frac{97}{100} \times \dots$$

until it falls below 0.5. At this number, the probability that two Arretians have the same birthday is greater than 0.5. The table similar to Table 5.1 is:

n	probability	n	probability	n	probability
2	0.9900	3	0.9702	4	0.9401
5	0.9035	6	0.8583	7	0.8068
8	0.7503	9	0.6903	10	0.6282
11	0.5659	12	0.5032	13	0.4428

The number of Arretians for a greater than 50% chance that two have the same birthday is 13.

5.2 The table similar to Table 5.2 is:

Outcome	Probability	Profit (loss) on first game	Profit (loss) on C&A
0 0	$\frac{5}{6} \times \frac{5}{6} = \frac{25}{36}$	(−25)	(−25)
0 6	$\frac{5}{6} \times \frac{1}{6} = \frac{5}{36}$	5	5
6 0	$\frac{1}{6} \times \frac{5}{6} = \frac{5}{36}$	5	5
6 6	$\frac{1}{6} \times \frac{1}{6} = \frac{1}{36}$	1	2
	Net loss	14	13

Problems 6

6.1 Consider a notional 200 patients with the symptoms. Of these, 130 will have condition A and 70 condition B. The number cured with:

drug a: $130 \times 0.6 + 70 \times 0.4 = 106$.
drug b: $130 \times 0.3 + 70 \times 0.9 = 102$.
Hence the doctor should prescribe drug a.

6.2 The O and E tables are:

	Buy	No buy
old	8	92
new	20	80

O

	Buy	No buy
old	14	86
new	14	86

E

$$\text{These give}: \quad \chi^2 = \frac{36}{14} + \frac{36}{86} + \frac{36}{14} + \frac{36}{86} = 5.98.$$

Table 6.1 indicates that the probability of getting this value or above is between 0.01 and 0.02. This is much less than the decision limit of 0.05. Hence the indication is that the new packaging is effective.

Problems 7

7.1 Each path from the top point to a cup has a probability of $\frac{1}{9}\left(\frac{1}{3} \times \frac{1}{3}\right)$. There is one path to A, so the probability of getting there (or to E) is $\frac{1}{9}$. There are two paths to B, so the probability of getting there (or to D) is $\frac{2}{9}$. There are three paths to C, so the probability of getting there is $\frac{1}{3}$. In 9 games, costing 90 p, the player receives 40 p from the ball reaching B and D, and 40 p from the ball reaching A or E, i.e., a total of 80 p in all.

7.2 The probabilities of the various ways of getting 20 are:

$$10 + 10 \quad \text{probability} \quad \tfrac{16}{52} \times \tfrac{15}{51} = \tfrac{240}{2652} = 0.09050$$

$$11 + 9 \quad \text{probability} \quad \tfrac{4}{52} \times \tfrac{4}{51} = \tfrac{16}{2652} = 0.00603$$

$$9 + 11 \quad \text{probability} \quad \tfrac{4}{52} \times \tfrac{4}{51} = \tfrac{16}{2652} = 0.00603$$

Hence the total probability is: 0.10256.

7.3 (i) The probability of setting a point of 4 and then losing by throwing a 7 is:

$$P_{4,7} = \text{Probability of getting 4}$$
$$\times \text{Probability of getting a 7 before a 4}$$
$$= \text{Probability of getting 4}$$
$$\times (1 - \text{Probability of getting a 4 before a 7})$$
$$= \frac{1}{12} \times \left(1 - \frac{1}{3}\right) = \frac{1}{18}.$$

(ii) With the notation of (i) we need to find $P_{4,7} + P_{5,7} + P_{6,7} + P_{8,7} + P_{9,7} + P_{10,7}$

$$P_{5,7} = \text{Probability of getting 5}$$
$$\times \text{Probability of getting a 7 before a 5}$$
$$= \text{Probability of getting 5}$$
$$\times (1 - \text{Probability of getting a 5 before a 7})$$
$$= \frac{1}{9} \times \left(1 - \frac{2}{5}\right) = \frac{1}{15}.$$

$$P_{6,7} = \text{Probability of getting 6} \times \text{Probability of}$$
$$\text{getting a 7 before a 6}$$

$$= \text{Probability of getting } 6 \times (1 - \text{Probability of getting a 6 before a 7})$$

$$= \frac{5}{36} \times \left(1 - \frac{5}{11}\right) = \frac{5}{66}.$$

We also have $P_{10,7} = P_{4,7}, P_{9,7} = P_{5,7}$ and $P_{8,7} = P_{6,7}$. Hence:

$$P_{4,7} + P_{5,7} + P_{6,7} + P_{8,7} + P_{9,7} + P_{10,7}$$

$$= \frac{1}{18} + \frac{1}{15} + \frac{5}{66} + \frac{5}{66} + \frac{1}{15} + \frac{1}{18} = 0.3960.$$

Problems 8

8.1 The expected number for each side is 100, so that:

$$\chi^2 = \frac{144}{100} + \frac{169}{100} + \frac{361}{100} + \frac{81}{100} + \frac{1}{100} + \frac{256}{100} = 10.12.$$

There are 5 degrees of freedom and according to Table 8.9 the probability of getting this value of χ^2, or some greater value, is between 0.05 and 0.1. This is less that the 10% criterion, so we conclude that the die is probably unfair.

8.2 The probability of getting either $1 + 6$ or $6 + 1$ is:
$2 \times \frac{1}{4} \times \frac{1}{12} = \frac{1}{24}$.

The probability of getting either $2 + 5$ or $5 + 2$ is:
$2 \times \frac{1}{6} \times \frac{1}{6} = \frac{1}{18}$.

The probability of getting either $3 + 4$ or $4 + 3$ is:
$2 \times \frac{1}{6} \times \frac{1}{6} = \frac{1}{18}$.

The probability of getting either $5 + 6$ or $6 + 5$ is:
$2 \times \frac{1}{6} \times \frac{1}{12} = \frac{1}{36}$.

Hence the probability of getting a natural is:

$$P_{natural} = \frac{1}{24} + \frac{1}{18} + \frac{1}{18} + \frac{1}{36} = 0.1806 \ (0.2222 \text{ for fair dice}).$$

8.3 The observed and expected tables are:

	0	non-0
Observed	20	350

	0	non-0
Expected	10	360

This gives $\chi^2 = \frac{100}{10} + \frac{100}{360} = 10.28$. For one degree of freedom the probability of getting this value of χ^2, or some greater value, is close to 0.001. Hence we may deduce that the wheel is almost certainly biased.

Problems 9

9.1 The histogram representing the distribution of dress sizes is shown below. If only sizes inside the range 8 to 18 were stocked, then the number of dresses sold would change by a factor 0.83. However, the profit per dress would change by a factor 1.2. The changed profit would be $0.83 \times 1.2 = 0.996$, so the total profit would fall.

Problems 10

10.1 The average is: $\dfrac{2+4+6+7+8+10+12}{7} = \dfrac{49}{7} = 7$

$$V = \dfrac{\begin{array}{c}(2-7)^2 + (4-7)^2 + (6-7)^2 + (7-7)^2 \\ + (8-7)^2 + (10-7)^2 + (12-7)^2\end{array}}{7} = 10$$

$$\sigma = \sqrt{10} = 3.162.$$

10.2 (i) 3^{11} (ii) 3^{12} (iii) $3^{1.3}$.

10.3 Sales of 47,000 are $2\frac{1}{2}\sigma$ below the mean. From Table 10.1 the area under the normal curve from the mean to $2\frac{1}{2}\sigma$ is 0.49379, so the area beyond $2\frac{1}{2}\sigma$ is 0.00621. Hence the number of days per year with sales below 47,000 will be: $0.00621 \times 365 = 2$, to the nearest whole number.

Sales of 55,000 are $1\frac{1}{2}\sigma$ from the mean. The area from the mean out to $1\frac{1}{2}\sigma$ is, from Table 10.1, 0.4332, so the area beyond $\frac{1}{2}\sigma$ is 0.0668. Hence the number of days per year with sales above 5,000 will be: $0.0668 \times 365 = 24$, to the nearest whole number.

Problems 11

11.1 numbers 2 4 6 7 8 10 12 mean $= 7$
squares 4 16 36 49 64 100 144 mean square $= 59$
Variance $=$ mean square $-$ square of mean $= 59 - 7^2 = 10$.

11.2 We first work out the mean square height for each school from $V + \bar{h}^2$.
School 1 $\overline{h^2} = 0.091^2 + 1.352^2 = 1.8362$
School 2 $\overline{h^2} = 0.086^2 + 1.267^2 = 1.6127$
School 3 $\overline{h^2} = 0.089^2 + 1.411^2 = 1.9988$
School 4 $\overline{h^2} = 0.090^2 + 1.372^2 = 1.8905$

The mean square height for all the children is

$$\overline{h_{all}^2} = \frac{\begin{array}{c}(62 \times 1.8362) + (47 \times 1.6127)\\ + (54 \times 1.9988) + (50 \times 1.8905)\end{array}}{62 + 47 + 54 + 50} = 1.8408$$

The mean height of all the children is

$$\overline{h_{a;;}} = \frac{\begin{array}{c}(62 \times 1.352) + (47 \times 1.267)\\ + (54 \times 1.411) + (50 \times 1.372)\end{array}}{62 + 47 + 54 + 50} = 1.353\,\text{m}$$

The variance of the heights for all the children is

$$V_{all} = \overline{h_{all}^2} - \overline{h_{all}}^2 = 1.8408 - 1.353^2 = 0.01019\,\text{m}^2$$

Hence the standard deviation is: $\sigma_{all} = \sqrt{V_{all}} = 0.1010\,\text{m}$.

Problems 12

12.1 For this Poisson distribution the average $a = 1$

(i) $\dfrac{e^{-1} \times 1^0}{0!} = 0.3679$ (ii) $\dfrac{e^{-1} \times 1^1}{1!} = 0.3679$

(iii) $\dfrac{e^{-1} \times 1^3}{3!} = 0.0613$ (iv) $\dfrac{e^{-1} \times 1^5}{5!} = 0.0031$

12.2 The average number of asteroids per 1,000,000 years is 0.1. The probability that no large asteroids will fall is: $\dfrac{e^{-0.1} \times 0.1^0}{0!} = 0.9048$. Hence the probability that at least one will fall is: $1.0 - 0.9048 = 0.0952$.

12.3 The number of days that 15 would fail is: $365 \times \dfrac{e^{-10} \times 10^{15}}{15!} = 12.7$, or 13 days to the nearest whole number.

Problems 13

13.1 The proportion of those polled supporting party A is 0.52. Hence the most probable proportional support for party A is

0.52 with a standard deviation:

$$\sigma = \sqrt{\frac{0.52 \times 0.48}{3000}} = 0.00912.$$

The indicated support is: $\frac{0.52-0.50}{0.00912}\sigma = 2.19\sigma$ from the 50% level. The area under the normal curve out to 2.19σ may be found from Table 10.1 by going $\frac{9}{10}$ of the way from 2.1σ towards 2.2σ. This area is 0.486. For party B to win, the actual result would have to be more than 2.19σ from the mean, the probability of which is: $0.5 - 0.486 = 0.014$. Hence the probability that party A will have more than 50% of the popular vote is: $1.0 - 0.014 = 0.986$.

13.2 From the sample the proportion of male snakes is 0.46. The standard deviation of this estimate is: $\sigma = \sqrt{\frac{0.46 \times 0.54}{200}} = 0.0352$. The sample proportion is: $\frac{0.5-0.46}{0.0352}\sigma = 1.14\sigma$ from the 50% level. For more than 50% of snakes to be male, the sample average must be more than 1.14σ from the mean. From Table 10.1, interpolating between 1.1σ and 1.2σ, the probability of this is: $0.5 - 0372 = 0.128$.

Problems 14

14.1 (i) The mean height is found by adding the given heights and dividing by 20. This gives the mean height as: $\bar{h} = 1.9615\,\text{m}$.

(ii) To find the variance we first need to find the mean square height of the men. This is found by adding the squares of their heights and dividing by 20. This gives: $\overline{h^2} = 3.8569\,\text{m}^2$. Hence the variance of the height is:

$$V_h = \overline{h^2} - \bar{h}^2 = 3.85694 - 1.9615^2\,\text{m}^2 = 0.0094577\,\text{m}^2$$

giving: $\sigma_h = \sqrt{V_h} = 0.097\,\text{m}$.

(iii) Applying the Bessel correction, the estimated standard deviation for the whole population is: $\langle \sigma_p \rangle = \sqrt{\frac{20}{19}} \times 0.097 = 0.0995$ m.

(iv) The estimated standard deviation of the sample mean is, from (14.4):

$$\langle \sigma_{\bar{a}_p} \rangle = \sqrt{\frac{V_h}{19}} = \sqrt{\frac{0.0094577}{19}} = 0.022 \text{ m}.$$

(v) 2.0 m is $\frac{2.0-1.9615}{0.022} \langle \sigma_{\bar{a}_p} \rangle = 1.75 \langle \sigma_{\bar{a}_p} \rangle$ from the sample mean. The probability of being this far, or further, from the sample mean is found from Table 10.1 as $0.5 - 0.459 = 0.041$.

14.2 The proportion of the sample tagged is $\frac{20}{100} = 0.2$. Assuming that this is true for all the fish then, since 100 fish were tagged, the total number of fish in the pond is 500.

Problem 15

15.1 We take the null hypothesis that there is no difference in the feeding regimes. The difference in mean weights from the two samples is 0.5 gm. The variance of the difference $V_{diff} = \frac{2.1^2 + 2.3^2}{100 + 100 - 2} = 0.04899$ gm^2 giving a standard deviation 0.22 gm. The actual difference is: $\frac{0.5}{0.22} = 2.27\sigma$. The probability of being that far, or further, from the mean in both directions is, from Table 10.1:

$$2 \times (0.5 - 0.4882) = 0.0236.$$

Problem 16

16.1 Estimating figures from a graph is not something that can be done precisely, but the following estimates are reasonable.

(i) Number killed in 1964   7,700
     Number killed in 2005   3,200
     Ratio = 0.42

(ii)   1964   1982   2005
        Pedestrians   3,000   1,800   850
        Total   7,700   5,950   3,200
        Ratio   0.39   0.30   0.30

The number of cars on the road in the period in question has greatly increased, which makes the reduction in road deaths more remarkable. The MOT test for cars was introduced in 1960 for cars over 10 years old, but the age at which the test was applied was reduced until, by 1967, it was 3 years. Cars have become much safer with the introduction of mandatory safety belts and airbags that protect both the driver and passengers. Front-wheel drive, power-assisted steering and improved suspension give better vehicle control, giving extra safety both to those in the car and to pedestrians. Advanced braking systems (ABS) have also contributed greatly to safety; enabling braking in shorter distances without skidding, particularly on icy roads.

Problem 17

17.1 (i)   0.095   (ii)   0.22   (iii)   0.24
    The exposure level is 2.0 without smoking and 4.0 with smoking. The probability of disease is increased from 0.175 to 0.28.

Problem 18

18.1 With $p_{10} = 0$, and hence $p_9 = 1$, the probability of a bad outcome by not invading is reduced to: $0.8 \times 0.7 + 0.2 \times 0 = 0.56$. This is now less than the probability of a bad outcome by invading.

Problem 19

19.1 The increase between 2000 and 2100 will be five times that between 1980 and 2000, i.e., 185 ppm for CO_2 concentration and $2.0°$ C for global anomaly. From Figure 19.3 in 2000, the CO_2

concentration was 368 ppm and the global anomaly 0.45° C. Hence the expected values in 2100 are:

CO_2 concentration $368 + 185 = 553$ ppm.
Global anomaly $0.45 + 2.0 = 2.45°$ C.

There will be the same increase between 2100 and 2200, so the expected values in 2200 are:

CO_2 concentration $368 + 370 = 738$ ppm.
Global anomaly $0.45 + 4.0 = 4.45°$ C.

If the emission of CO_2 is halved after 2020, then the increase of both CO_2 concentration and global anomaly between 2000 and 2100 will be just three times that between 1980 and 2000, i.e., 111 ppm for CO_2 concentration and $1.2°$ C for global anomaly. In this case, the expected values in 2100 are:

CO_2 concentration $368 + 111 = 479$ ppm.
Global anomaly $0.45 + 1.2 = 1.65°$ C.

Between 2100 and 2200 the increase will be 2.5 times the increase between 1980 and 2000. Hence the expected values in 2200 are:

CO_2 concentration $479 + 2.5 \times 37 = 571.5$ ppm.
Global anomaly $1.65 + 2.5 \times 0.4 = 1.65°$ C.

Problem 20

20.1 Raising the retirement age to 66 increases the percentage of those contributing by 0.88 (the estimated number between ages 65 and 66) and reduces those receiving pensions by the same amount. Hence we now have:

$$T = 59.71CfI \quad P = 14.99CpA \quad \text{and} \quad L = 0.88CA$$

with $f = 0.22$ and $p = 0.475$ and other values as before.

The total income to the fund is:

$$(1 - s)T = 0.95 \times 59.71 \times 0.22 \times 20{,}000C = 249{,}588C.$$

The total outgoings are:

$$P + L = CA(14.99\,p + 0.88) = 30{,}000 \times (14.99 \times 0.475$$
$$+ 0.88)\,C = 240{,}007C.$$

Hence the fund now has a small annual surplus of 9581C.

Index